Pioneering the Future for Traffic and Safety Sciences

Pioneering the Future for Traffic and Safety Sciences

Transdisciplinary Wisdom

The International Association of Traffic
and Safety Sciences (IATSS)
Tokyo, Japan

ISBN 978-981-96-0675-7   ISBN 978-981-96-0676-4 (eBook)
https://doi.org/10.1007/978-981-96-0676-4

The original submitted manuscript has been translated into English. The translation was done using artificial intelligence. A subsequent revision was performed by the author(s) to further refine the work and to ensure that the translation is appropriate concerning content and scientific correctness. It may, however, read stylistically different from a conventional translation.
Translation from the Japanese language edition: "Traffic and Safety Sciences" by The International Association of Traffic and Safety Sciences (IATSS), © The International Association of Traffic and Safety Sciences (IATSS) 2024. Published by Maruzen Publishing Co., Ltd.. All Rights Reserved.

© The International Association of Traffic and Safety Sciences (IATSS) 2026. This book is an open access publication.

Open Access This book is licensed under the terms of the Creative Commons Attribution-NonCommercial-NoDerivatives 4.0 International License (http://creativecommons.org/licenses/by-nc-nd/4.0/), which permits any noncommercial use, sharing, distribution and reproduction in any medium or format, as long as you give appropriate credit to the original author(s) and the source, provide a link to the Creative Commons license and indicate if you modified the licensed material. You do not have permission under this license to share adapted material derived from this book or parts of it.
The images or other third party material in this book are included in the book's Creative Commons license, unless indicated otherwise in a credit line to the material. If material is not included in the book's Creative Commons license and your intended use is not permitted by statutory regulation or exceeds the permitted use, you will need to obtain permission directly from the copyright holder.
This work is subject to copyright. All commercial rights are reserved by the author(s), whether the whole or part of the material is concerned, specifically the rights of reprinting, reuse of illustrations, recitation, broadcasting, reproduction on microfilms or in any other physical way, and transmission or information storage and retrieval, electronic adaptation, computer software, or by similar or dissimilar methodology now known or hereafter developed. Regarding these commercial rights a non-exclusive license has been granted to the publisher.
The use of general descriptive names, registered names, trademarks, service marks, etc. in this publication does not imply, even in the absence of a specific statement, that such names are exempt from the relevant protective laws and regulations and therefore free for general use.
The publisher, the authors and the editors are safe to assume that the advice and information in this book are believed to be true and accurate at the date of publication. Neither the publisher nor the authors or the editors give a warranty, expressed or implied, with respect to the material contained herein or for any errors or omissions that may have been made. The publisher remains neutral with regard to jurisdictional claims in published maps and institutional affiliations.

This Springer imprint is published by the registered company Springer Nature Singapore Pte Ltd.
The registered company address is: 152 Beach Road, #21-01/04 Gateway East, Singapore 189721, Singapore

If disposing of this product, please recycle the paper.

Features and Utilization of This Book

The title of this book, *Pioneering the Future for Traffic and Safety Sciences*, aims to broadly capture the various issues surrounding our transportation society, compile the wisdom of various experts on the technical systems and social systems that support traffic safety, and make proposals for the future.

This book consists of 13 chapters, each of which highlights keywords that each expert considers essential for learning about the next generation of transportation society and adds explanations. Therefore, it is not necessary to start reading from the first chapter, and it is designed to be understood even if you read across chapters and sections. The flow of this book is outlined as follows.

Chapters 1 and 2 feature urban engineering experts discussing the nature of cities and transportation, and explaining the knowledge necessary to build the future transportation society. In Chap. 3, an environmental science expert outlines the relationship between environmental issues and transportation, and provides insights toward a low-carbon society that also considers ecosystems.

From Chap. 4 onward, individual discussions are developed. Chapter 4 explains the relationship between public transportation and urban and regional activities, and explains the latest technologies in public transportation, including DX. Chapter 5 explains the basics of traffic engineering, such as traffic flow, the mechanism of traffic accidents, and road structure.

Chapters 6, 7 and 8 focus on technologies for improving safety, from both the infrastructure and automobile perspectives. Chapter 6 explains safety technologies from the perspective of road infrastructure, such as intersection structures. Chapter 7 explains safety technologies from the perspective of vehicles, such as vehicle safety and ASV (Advanced Safety Vehicle). Chapter 8, in particular, is a standalone chapter on autonomous driving, a major theme in recent years, providing an overview of technology development and international standardization activities.

Chapters 9, 10 and 11 explain traffic safety from the human perspective, such as drivers and traffic participants. Chapter 9 explains traffic psychology, such as traffic behavior models and risk perception. Chapter 10, as traffic education, explains traffic safety education as civic education and training models for drivers. Chapter 11 explains emergency medical care for traffic injuries and accidents related to driver health.

Chapter 12 discusses legal issues related to traffic safety, including problems with traffic accidents and legal issues with autonomous driving, a rapidly developing theme in recent years. Chapter 13 explains traffic policy, congestion, and environmental issues from the perspective of behavioral economics and cost-benefit analysis.

Finally, under the title "Contribution to Sustainable Transportation Society and SDGs," the positioning of the transportation sector for the 17 areas of the SDGs is discussed. Also, the development of "Transportation and Safety Studies" is overviewed in the context of the 50-year history of IATSS.

In this way, this book is intended to enable readers to acquire a multifaceted perspective on "Traffic and Safety Studies" by providing a cross-sectional explanation of each field. This book is a work that strongly recognizes the "interdisciplinarity" characteristic of the International Traffic Safety Association, and it would be fortunate if the readers could discover new insights buried between each chapter and section by freely interpreting them.

Introduction

In order to enhance today's transportation society into a safer and more sustainable one, a diverse collection of knowledge spanning from the natural sciences to the social sciences and the humanities is required. Moreover, even in the global trend of motorization, there are significant disparities in daily mobility and accessibility to life opportunities between developed and developing countries, and between large cities and rural areas, suggesting the importance of an approach that values the diversity of transportation society. In this context, what we, the International Association for Traffic and Safety Sciences, are called upon to do is to construct "traffic and safety studies" that pursue interdisciplinary and practical knowledge based on scientific evidence and local circumstances.

Under this philosophy, as part of the 40th anniversary project of the International Association for Traffic and Safety Sciences in 2014, *Traffic and Safety Studies* was published in 2015. In 2024, as part of the 50th anniversary project of the International Association for Traffic and Safety Sciences, *Traffic and Safety Studies for the Future* was published. This book is a revised version of the previously published *Traffic and Safety Studies*, incorporating developments and insights in the field over the past 10 years, as well as proposals for the future. The writing and editing of this book were carried out mainly by members of the International Association for Traffic and Safety Sciences. The authors represent traffic engineering, urban engineering, electrical engineering, information engineering, system engineering, mechanical engineering, environmental studies, psychology, medicine, law, administrative studies, economics, and sustainability studies, interconnecting the essence of knowledge in each field. In recent years, the environment for actively discussing themes such as DX (Digital Transformation) and AI (Artificial Intelligence) in the field of transportation has been established. Therefore, this book includes expanded research themes that have rapidly developed over the past 10 years, such as public transportation and MaaS (Mobility as a Service), technology and law of autonomous driving, SDGs (Sustainable Development Goals), and sustainability. It has been decided to publish this book not only in Japanese but also in English. The latter is intended to be used for education and research in developing countries in the Asian region.

The *Traffic and Safety Studies* published in 2015 was composed of two parts: theory and practice. However, this book, *Pioneering the Future for Traffic and Safety Sciences*, is composed only of chapters equivalent to the theoretical part. As something equivalent to the practical part of this book, the English journal published by our association, *IASS Research*, has planned a special issue "History and Social Impact of IATSS Research Projects." This is a special issue composed of papers written by each project leader about the results of research projects conducted by the International Association for Traffic and Safety Sciences so far. This project is also part of the 50th anniversary project of the International Association for Traffic and Safety Sciences, and many of the authors of the special issue papers are also authors of this book. The special issue papers are scheduled to be published sequentially in issues 48-2 and 48-3 of IATSS Research. We hope you will read this book along with these articles.

Finally, I would like to express my heartfelt gratitude to all the authors and related parties involved in the planning and writing of this book.

Contents

1. **Urban and Transportation** 1
   Masanobu Kii

2. **Land Use and Transportation** 15
   Akinori Morimoto

3. **Transportation and the Environment** 31
   Tomohiro Ichinose

4. **Public Transportation and the Local Community** 53
   Fumihiko Nakamura, Ryo Ariyoshi, and Yurie Toyama

5. **Traffic Engineering** 77
   Takashi Oguchi, Miho Iryo, and Azusa Toriumi

6. **Technology for Improving Safety: Infrastructure Part** 103
   Koji Suzuki and Hideki Nakamura

7. **Technology for Safety Improvement: Automobile Edition** 119
   Taro Sekine

8. **Automated Driving and Driver Assistance** 141
   Yoichi Sugimoto and Kunimichi Hatano

9. **Traffic Psychology** 163
   Kazumitsu Shinohara and Kazuko Okamura

10. **Traffic Education** 183
    Yuto Kitamura, Kazuhisa Ogawa, and Nagahiro Yoshida

11. **Traffic Safety and Medicine** 199
    Kazuhiko Kibayashi, Takashi Moriya, Migiwa Asano,
    and Masaya Takahashi

12	**Legal System to Ensure Traffic Safety**	219
	Takeyoshi Imai	
13	**Sustainable Growth: An Economic Perspective**	235
	Kazuhiro Ohta, Mariko Futamura, and Akihiro Nakamura	

Epilogue 255

Chapter 1
Urban and Transportation

Masanobu Kii

1.1 Emergence of Transportation Issues

Urban transportation issues have been recognized since the 1920s when motorization advanced in the United States. An architect Harvey W. Corbett proposed solution to traffic congestion in urban centers in 1950 in the *Popular Science Monthly* in 1925, subtitled "buildings half-mile high and 4-deck streets may solve congestion problems."

The upper floors of the buildings alongside the central street are used for residences and parks, the middle floors are schools, the lower floors are offices, and the first and second floors are restaurants. In other words, it is depicted that daily life is completed within a huge building.

Also, at the central street, the surface is a pedestrian-only space, the basement first floor is a passage for slow-moving cars. The basement second floor is a passage for fast-moving cars, and the basement third floor is a railway line. This prediction for 1950 presented by an architect in 1925 has been partially realized in many major cities around the world today.

In the explanation of this proposal, it was assumed that by 1950, infrastructure would solve the congestion problems of cities, and a bright future was expected due to technological progress. Toward this bright future, a vast amount of transportation infrastructure has been developed, but, by the 1960s, it was recognized that congestion problems could not be solved by infrastructure development alone. This is because when congested roads are widened and convenience is improved, the number of users increases accordingly, creating induced demand. To consider induced demand at the planning stage, it is necessary to understand how much infrastructure development increases the demand for movement and, in the long term, how much it brings about the location of housing and industry.

M. Kii (✉)
Graduate School of Engineering, The University of Osaka, Osaka, Japan
e-mail: kii@see.eng.osaka-u.ac.jp

Not only in the United States but also in the other countries, congestion, environment, and safety have continued to be important issues in urban transportation after motorization. On the other hand, in recent years in Japan, population decline has begun even in large cities, and the maintenance of transportation services and infrastructure has become an issue. In addition, revitalization of the region through low-speed movement use and high-quality movement space and health promotion through active transport are also themes of urban transportation policy. The intensity of these issues varies from city to city, but one of the main drivers that generate various urban transportation issues is the city's population. Generally, congestion is more serious in populous cities, and the maintenance of public transportation, which is an issue in many small- and medium-sized cities, is affected by aging and population decline. Urban transportation issues vary depending on the size and composition of the city's population, and therefore, the methods of solving these issues are not necessarily uniform.

In this chapter, we first discuss the distribution of urban population size so far and the mechanism of migration. Next, we consider the size of cities, transportation services, and living standards. Based on this, we summarize the transportation issues of large cities and small- and medium-sized cities and look ahead to the direction of problem solving by technology and policy.

1.2 Distribution of Urban Population Size

The distribution of urban population size is empirically said to follow a power law. This is a law that the population size of the kth city is proportional to the population of the first city to the power of k^{-s} (s is a constant). Under this law, the relationship between the logarithm of urban population size and the logarithm of rank becomes a straight line. Figure 1.1 shows the population size and rank for cities with a population of 300,000 or more from 1950 to 2010, with the logarithm of population on the horizontal axis and the logarithm of rank on the vertical axis, for six countries.

From this, there is variation by country, but in any country, it can be seen that the distribution of urban population size generally follows the power law. This distribution indicates that there are few cities of large scale, and as the scale decreases, the number of cities increases exponentially. Also, the power law holds when there are agglomeration effects in cities [1]. That is, such a distribution is observed when the population tends to increase more in large cities.

However, the trend of annual changes in distribution varies by country. Except for the UK, the urban population has increased in all countries during the target period, and the plot as a whole is shifting to the right. France and Thailand show an extreme concentration type, with the slope of the line segment connecting the largest city and the second city being significantly gentler than the line segment after the

1 Urban and Transportation 3

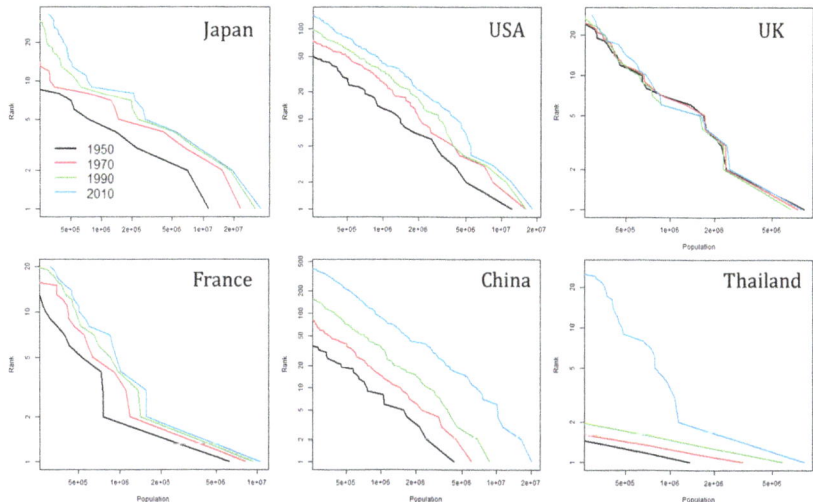

Fig. 1.1 Distribution of urban population size by country

second city. Similarly, the UK also has a high degree of concentration. However, in France and Thailand, the entire plot is shifting to the right, and it can be seen that the population of the second city and beyond is also increasing along with the largest city. Japan, the United States, and China have a dispersed urban size distribution compared to the UK, France, and Thailand. However, looking in detail, in the United States and China, the slope of the top plot (right end) is increasing over time, indicating a trend toward increasing dispersion of urban size, while in Japan, the slope has become gentler from 1970 to 2010, indicating an increasing degree of population concentration in the metropolitan area. Also, in Japan, it appears that the population of medium-sized cities is increasing, but this also reflects the impact of municipal mergers.

Figure 1.2 shows the population of cities around the world in 2010 and projected values for 2100 [2]. From these results, it is expected that the urban population will increase significantly, especially in developing countries from South Asia to Africa, and both large cities and small- and medium-sized cities are expected to increase significantly. The projected future urban size distribution also follows the power law.

From these observations and future projections, it is understood that cities do not converge to a specific population size but distribute according to the power law. However, the distribution shape that represents the degree of population concentration in large cities and dispersion in small- and medium-sized cities varies by country. In the next section, we will consider the reasons for the differences in city size and their relationship with transportation.

(a) 2010

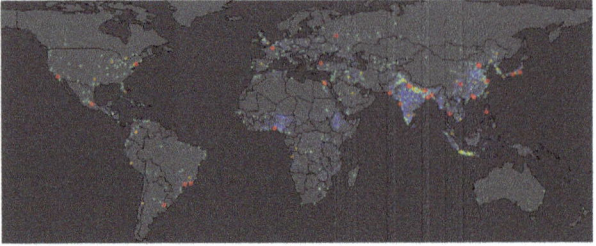

(b) 2100, Shared Socioeconomic Pathway scenario 1

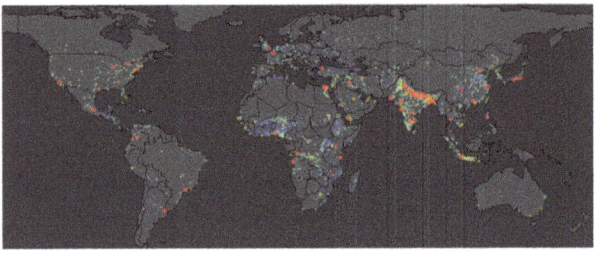

● 10million- ● 5-10million ● 1-5million ● 0.5-1million · 100-500 thousand

Fig. 1.2 Spatial distribution of urban population in 2010 (**a**) and 2100 (**b**) [2]

1.3 Agglomeration, Congestion Effects, and Intercity Population Movement

The factors that determine changes in urban population include natural increase and decrease due to differences in birth and death rates, as well as social change due to intercity migration. In particular, the trend of concentration in the largest city, seen in several countries including Japan, is due to the migration. Based on Kanemoto [3], we will schematically show the mechanism of intercity migration and the effect of regional policies.

First, it is assumed that people can freely migrate between cities and choose the city with high utility as their place of residence. Utility is assumed to be a comprehensive living standard index composed of income, living cost, transportation burden, and amenities derived from nature and culture when living in that city. For simplicity, we will consider migration and changes in utility for two cities.

Figure 1.3 assumes two cities A and B and shows the utility of cities A and B u_A, u_B in relation to the population of city A N_A. Here, the total population of the two cities N is assumed to be constant, and the population of city B $N_B = N - N_A$ is represented by the distance from the left end of the graph being N_A and the distance from the right end being N_B. Here, it is assumed that cities have effects of agglomeration and congestion. If the population is small, there is no congestion, but

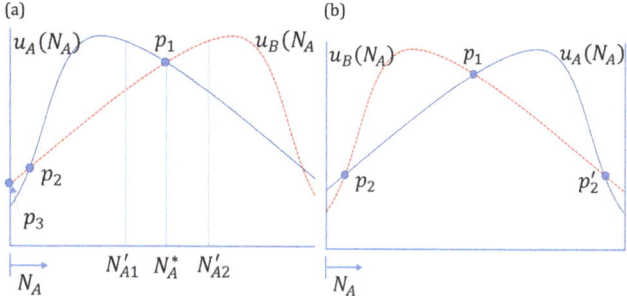

Fig. 1.3 Utility curves of two cities (**a**) case where congestion effect is predominant with equal population, (**b**) case where agglomeration effect is predominant with equal population)

productivity is low, so income is low, and it is assumed that the utility level is low. As the population increases, productivity improves and income increases, so utility increases. However, it is assumed that when the population exceeds a certain level, the effect of congestion exceeds the agglomeration effect and utility decreases. Here, it is assumed that the agglomeration effect means that productivity and some amenities increase as the city population increases, and the congestion effect includes increases in living costs such as traffic congestion and housing costs.

In this way, it can be considered that the utility with respect to the urban population is represented by a curve with a maximum value. Figure 1.3a represents the case where the utility curve $u_A(N_A)$ peaks when $N_A < N/2$. This means that the agglomeration effect manifests at a stage where the population is small, but the effect of congestion becomes dominant when $N_A < N/2$. At that time, points p_1 and p_2 become equilibrium points where population migration does not occur because the utilities of the two cities are equal. Also, p_3 is a point where population migration does not occur anymore, even though the utilities are not equal, because the population of the city with lower utility is zero. Here, p_1 and p_3 is a stable equilibrium point, whereas, p_2 is an unstable equilibrium point. For instance, p_1 shifted to the left to N'_{A1} where, $u_A > u_B$, the population moves from B to A, and eventually, the population of A becomes equal to the utility at N^*_A. The population also moves in the same dynamics until it becomes N^*_A in the case of N'_{A2}. p_3 Therefore, even if the population of A is slightly more than zero, $u_A < u_B$, the population of A will migrate to B and become zero. On the other hand, if the population of A is slightly more than p_2, $u_A > u_B$, the population of A will move to B until it becomes N^*_A. Also, if the population of A is slightly less than p_2, $u_A < u_B$, the population of A will migrate to B until it becomes zero. That is, when the population changes slightly, the population will move away from p_2 in p_2, whereas, p_1 and p_3 return to these points.

In Fig. 1.3a, if the effects of accumulation and congestion are the same in two cities, it is a stable equilibrium for the population to be the same in both cities, and the level of utility is higher than the equilibrium point where all the population is concentrated in one city p_3. Figure 1.3b shows the same utility curve as (a), but the city population that provides the maximum utility is different. In the figure, the utility curve $u_A(N_A)$ peaks at $N_A > N/2$, and while the effect of accumulation is dominant at a population smaller than the peak, the impact of congestion increases sharply at a population larger than that, resulting in a significant decrease in utility.

In this figure, contrary to Fig. 1.3a, p_2 and p_2' become stable equilibria and will increase to p_1. Therefore, p_1 is an unstable equilibrium. Specifically, if N_A is smaller than p_1, then $u_A < u_B$, so N_A will decrease to p_2. On the other hand, if N_A is smaller than p_2, then $u_A > u_B$, so N_A increases up to p_2. Also, if N_A is larger than p_1, N_A increases up to p_2'. This graph represents the situation of population migration from small and medium cities to large cities. At p_2, productivity is low in small cities, congestion worsens in large cities, and as a result, the utility level is lower than the unstable equilibrium point where the population of both cities is equal to p_1. p_2 can be interpreted as representing the state of overcrowding and depopulation of the two cities.

In the case of Fig. 1.3b, the equilibrium utility level rises due to income transfer from large cities to small cities. Figure 1.4 shows the utility curve when income is transferred from city A to city B. The thin line represents the utility before income transfer, and the thick line represents the utility after income transfer. Since income is transferred out from A, u_A is pushed down. On the other hand, since B receives income, u_B is pushed up. Then, the equilibrium point changes from p_2 to p_2', and the equilibrium utility increases. In other words, when the equilibrium utility level is low due to overcrowding or depopulation, income transfers through grants, or public works can be justified from the perspective of economic efficiency.

However, the conclusion is different if the disutility of congestion is smaller than the effect of agglomeration. Figure 1.5 shows a case where concentrating all the population in one city results in the highest utility. This is different from Fig. 1.3b in that the agglomeration effect outweighs the disutility of congestion, so there is no peak in the utility curve, and utility increases uniformly with respect to the population. Therefore, the equal population p_1 is an unstable equilibrium, and if N_A is larger, it reaches p_2', and its utility level is higher than p_1.

Such discussions of urban economics in textbooks are abstract and omit many important issues related to urban activities. Also, the conclusions can vary greatly depending on the assumed conditions. However, they are rich in suggestions for considering measures to deal with urban transportation problems in the post-corona era.

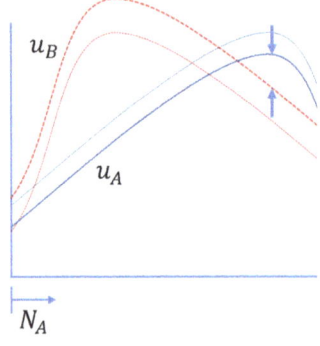

Fig. 1.4 Changes in equilibrium utility levels due to income transfers between cities

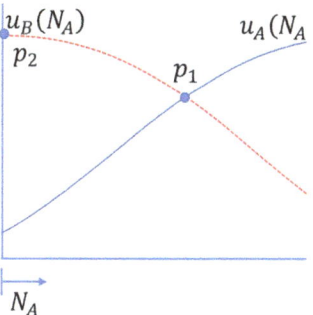

Fig. 1.5 Utility curve in cases where the effects of agglomeration are high and congestion disutility is low

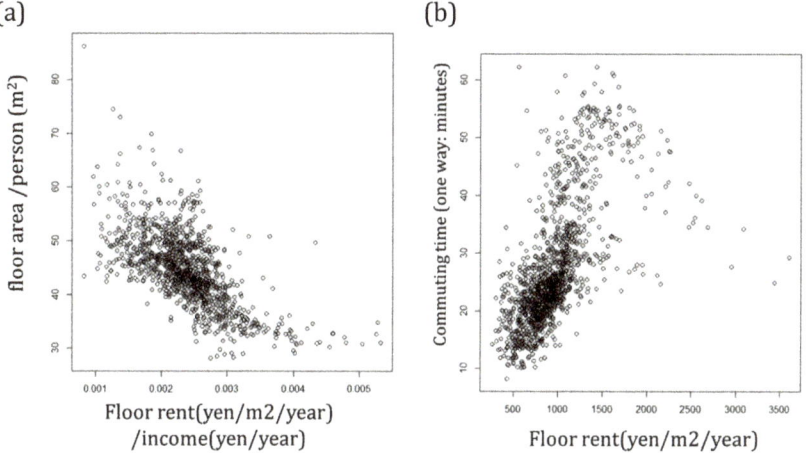

Fig. 1.6 Relationship between housing rent, housing area, and commuting time. (Created based on the 2018 Housing and Land Survey)

1.4 Housing and Commuting Time

The previous section theoretically demonstrated the mechanism by which differences in urban scale arise. In this section, based on statistics, we will show the situation of housing and commuting that results from these differences in urban scale.

Figure 1.6a shows the relationship between housing rent/income and per capita housing area in each municipality in Japan in 2018. The horizontal axis represents the average annual housing rent per floor area relative to per capita annual income, and the vertical axis represents the average housing area per person. From this, it can be seen that the higher the rent relative to income, the smaller the housing area per person. Generally, the larger the city, the higher the income, but on the other hand, the higher the housing rent. This graph shows that the rate of increase in housing rent relative to urban scale is higher than the rate of increase in income, and the larger the urban scale, the smaller the housing area per person.

Figure 1.6b shows the relationship between housing rent and commuting time. From this figure, a positive correlation can be seen between housing rent and commuting time in areas where the housing rent is less than 1800 yen/m2/year, and a negative correlation can be read in areas where it is higher. The former indicates a tendency for rents to increase and commuting times to lengthen as the size of the city increases. On the other hand, the latter represents a situation where the closer you get to the city center in a large metropolitan area, the higher the rent and the shorter the commuting time.

From these figures, it can be seen that as the size of the city increases, the average living space per person decreases, and commuting time increases. Also, within large metropolitan areas, the closer one gets to the city center, the shorter the commuting time, but as a trade-off, the rent increases, and the affordable living space decreases. These statistics probably align with our experiences. At the same time, these results indicate that while income is higher in larger cities, the utility based on consumption under time and income constraints is not necessarily high.

Generally, if the rent is high, the income of the landowner also increases. Therefore, if you own real estate in a large city, as the urban population increases, the level of utility will likely increase due to the income rise associated with the effects of agglomeration. On the other hand, if you live in a rented house without owning land, or if you buy a new house, the larger the city, the larger the housing expenditure, and the utility may decrease. Looking at the domestic migration situation before the Corona disaster, the population inflow to large cities, especially the metropolitan area, continued. Looking at the age composition of migrants, there are many young people in their 20 s, who move to large cities for opportunities such as university advancement and employment. These migrants often live in rented houses or buy new homes, and according to the analysis mentioned above, the residential utility based on consumption may be lower than in regional cities. Nevertheless, according to a survey by the Ministry of Land, Infrastructure, Transport and Tourism, reasons for migration include "employment in the desired occupation," "wages and other conditions," and "desired place of study" [4], and it is presumed that the agglomeration of industries and universities is the main factor for migration to urban areas.

Of course, the richness of life is not determined solely by living space and commuting time but is influenced by a more diverse set of components, and what is considered desirable varies from person to person. However, if issues related to housing costs and traffic congestion are causing a decline in the quality of life, policies to curb population concentration in large cities may be justified. Regional revitalization measures aimed at creating employment and educational opportunities outside of large cities could also be effective as a congestion countermeasure in large cities.

Next, as an example of a large city, we will show the traffic situation in the Tokyo metropolitan area. Figure 1.7 shows the concentrated traffic volume per area for outgoing and returning purposes based on the 2018 Tokyo Metropolitan Area Person Trip Survey data. From this, it can be seen that both are along the railway lines from the center, but the peak of concentrated traffic density is higher for outgoing

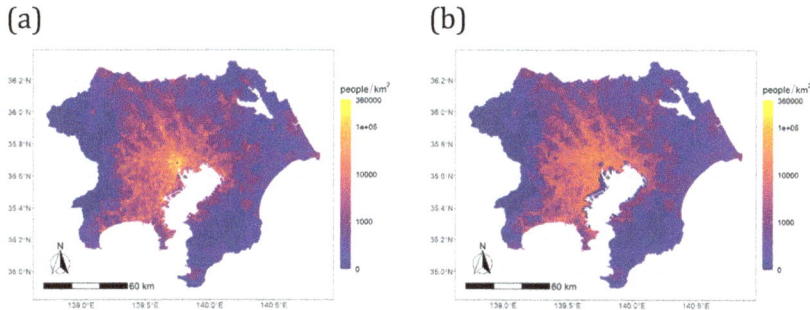

Fig. 1.7 Concentrated traffic volume (**a**: outgoing purpose, **b**: returning home purpose). (Created based on the Tokyo Metropolitan Area Person Trip Survey (2018). https://nlftp.mlit.go.jp/ksj/gml/datalist/KsjTmplt-S05-a-2013.html. Created using "R" with "simple feature" and "ggplot2" packages)

purposes than for returning purposes. Also, while the density in the center is high for outgoing purposes, it can be seen that there are many peaks in the suburbs as well. From this, it can be seen that the Tokyo metropolitan area has a polycentric urban structure, and the hierarchy of the urban structure can be further understood [5].

Many of the core functions of industry and administration are located in the central part of the metropolitan area, and the workforce that supports them is forced to choose between paying high housing rents, purchase costs, and spending a long-time commuting. According to the previous figure, in large metropolitan areas, housing costs are higher and commuting times are longer on average compared to small- and medium-sized cities. The high productivity and income distribution obtained through the effects of agglomeration are overshadowed by land constraints, resulting in smaller living spaces. Also, in order to make high use of land, housing becomes high-rise, and ultimately, high housing costs become land rent and capital costs, resulting in income for the limited landowners and capitalists. On the other hand, if technological progress or changes in production styles relax the spatial constraints necessary to express the effects of agglomeration, this could also relax land constraints and enhance the utility of residents. As mentioned later, the development of ICT technology and the corresponding changes in production styles may improve the quality of life through urban space and transportation activities.

1.5 Challenges and Directions for Solving Urban Transportation

As mentioned in Sect. 1.1, transportation challenges include congestion, environment, safety, spatial quality, regional vitality, health, and more. These challenges are interrelated, and the situation of the challenges varies depending on the size and structure of the city, and the measures to respond to these challenges also differ. Advanced cities tend to attract attention from the scale of measures, but various initiatives are being taken in small- and medium-sized cities as well. A comprehensive organization of these challenges and measures is left to another chapter, and this chapter discusses the challenges of traffic congestion in large cities and the maintenance of transportation services in small- and medium-sized cities and discusses the direction of their solutions.

The transportation challenges of both large cities and small- and medium-sized cities are affected by population dynamics. In light of the discussion in the previous section, the traffic congestion in large cities is a challenge of congestion disutility that exceeds the effects of agglomeration, and the decline in public transportation services in small- and medium-sized cities can be seen as a challenge of business efficiency decline due to a decrease in users, that is, a decrease in the effects of agglomeration.

First, in response to the issue of traffic congestion in large cities, traditionally, the expansion of transport capacity has been pursued as a basic measure. In particular, the enhancement of public transportation capacity during peak times has been addressed as an issue not only in transportation business but also in transportation administration, and the public transportation network in large urban areas has been expanded. This has contributed to alleviating congestion during peak times and has reduced the congestion rate of railways [6]. Convenient public transportation is essential for supporting economic activities in large cities, and its investment should continue. However, increasing transport capacity, as shown in Fig. 1.3, shifts the congestion curve upwards. In this case, if the situation is as in Fig. 1.3b and the productivity of small cities does not increase, in the long term, the population will move from small cities to large cities until the utility level balances at a point where congestion worsens.

Due to the state of emergency declaration in the COVID-19 pandemic, 38% of people have experienced telework, and 66% of them wish to continue [7]. In the survey, over 80% of respondents cited "reduction in commuting time" as a benefit of telework, the highest response. Commuting time is outside of working hours and is not included in production costs. Also, mental and physical fatigue due to congestion during commuting is usually not converted into economic costs. Therefore, unless the cost of introducing telework, such as changes in business processes and the establishment of remote systems, is less than the savings in commuting costs and office rents, there is little significance for employers to introduce telework. According to a 2022 survey by the Tokyo Metropolitan Government [8], about three-quarters of companies have introduced telework, and they report high effectiveness in reducing employees' commuting time, work-related travel time, and in dealing with employees who are raising children or caring for the elderly. Therefore, if the cost of securing personnel for companies decreases due to the reduction in commuting, which is a cost for workers, telework is likely to become more widespread.

So, how would telework affect utility in Fig. 1.3? The reduction in generalized commuting costs, including time and fatigue, corresponds to a reduction in congestion-related costs, so the utility of the congestion area in the figure would shift upwards. On the other hand, if telework can substitute the agglomeration effect in physical space in the information space, productivity may increase even with less population concentration, and utility outside the congestion area is expected to increase. However, there are tasks and public goods that cannot be substituted online, so the effect of agglomeration will not completely disappear, and there will also be services that require face-to-face interaction, so congestion will not completely disappear. Ultimately, depending on the degree of substitution of activities in physical space in the information space, telework is thought to both promote and alleviate the concentration of population in large cities, changing the shape of the utility curve in Fig. 1.3. However, both of these can reduce commuting traffic demand within large cities and potentially contribute to congestion relief.

It should be noted that many studies suggest that communication and transportation not only substitute for each other but also complement each other [9, 10] Activities in the information space can certainly induce activities in the real space. However, unless activities in the real space are concentrated in time and space, they are unlikely to exacerbate congestion. It is necessary to clarify the generation mechanism of complementary traffic demand and to develop spatio-temporal guidance measures.

In small- and medium-sized cities, the decrease in users makes it difficult to maintain public transportation services, and people who cannot use cars are facing the problem of reduced mobility. The decrease in the number of public transportation users is influenced by the decrease in population and the increase in the share of cars. Public transportation services are limited to areas where a certain demand is expected, but cars can be used as long as there are roads, even if the demand is small, because the users themselves prepare the means of transportation. For this reason, people can live on cheap land with low usage density when cars become available. When people use cars and move from the supply area of public transportation to the non-supply area, the number of public transportation users naturally decreases. If such a choice of residence is a rational choice, there may be no social problem. However, due to the incompleteness of foresight regarding the possibility of using cars when people become elderly, and the externality of road infrastructure and environmental load, cities dependent on cars may have inefficient spatial structures. We will examine these in more detail.

First, in the choice of residence, it may not be fully considered that driving a car will become difficult in the future due to aging or illness, or the possibility of selling the land in the future. If you live in an area where there is no means of transportation other than a car, even if your driving ability decreases due to aging, you will have to drive yourself if you cannot move to a convenient district, which will increase the risk of traffic accidents. Alternatively, it may be necessary to increase the public burden on public transportation services from a welfare perspective in order to guarantee a minimum cultural life. These can be considered as social costs due to low agglomeration.

Road infrastructure is a public good, and it does not become obsolete or increase the cost to users just because the number of users is small, but the cost of maintenance and management is usually borne widely by society. On the other hand, many public transportation routes also receive internal subsidies from operators and public subsidies from the perspective of securing the means of transportation for the transportation disadvantaged. However, the public transport is basically a private good and will be abolished if it does not establish as a private business. In this way, the cost burden structure is different between roads and public transportation. Also, because the cost of environmental load is not internalized, cars may be overused. Reflecting these, in small- and medium-sized cities, the decrease in population and the low density due to suburban development are reducing the efficiency of public transportation. If the city area is reduced and densified according to the decrease in population, it is expected that the activity efficiency of people who do not use cars can be improved.

The Compact Plus Network, a key initiative of the Ministry of Land, Infrastructure, Transport and Tourism, aims to create towns where not only car users but also others can live comfortably, by guiding urban facilities to hubs and reconstructing public transportation networks in coordination with them. If residential density increases, the efficiency of infrastructure and transportation services will also increase, and furthermore, productivity is likely to increase as the travel distance decreases and the supply efficiency of various services increases. However, the "Location Normalization Plan" system for guiding urban facilities is a mild system based on budget measures for promoting location and notification of development and does not necessarily guarantee the realization of facility development. Normally, if the density of land use is increased, the land price and location cost will be higher. In order to encourage the location of urban facilities and housing in the promoted area, it is necessary to have merits that match the cost.

Improving the quality of the living environment and convenience may attract residents even if the cost is high. For example, in the Marugame-machi shopping street in Takamatsu City, housing, medical/nursery facilities, and high-quality pedestrian spaces have been developed in conjunction with the redevelopment of the shopping street, and despite the high location cost compared to the suburbs, they have been successful in attracting residents. Location promotion requires a design of cost-effectiveness for the people and firms.

In addition, the measures aiming at internalizing the aforementioned externalities are effective, and the smart city technologies will substantially enhance the impacts. For example, if the efficiency of infrastructure and transportation services can be captured spatially by sensors or users' location information, such information can be used not only for administrative purposes such as urban infrastructure development and transportation measures, but also for private economic activities such as residential land development and consideration of new transportation services. Furthermore, the spatial distribution of property tax revenues reflecting infrastructure costs and land use value can serve as evidence for supporting urban compaction measures. Moreover, the obtained data can also be used for modeling the effects to analyze the policy measures. These will enhance the rationality of urban policies and contribute to the realizing the compact urban form of small- and medium-sized cities.

In conclusion, we have discussed specific issues for both large and small cities and examined directions for their resolution. Both are caused in part by changes in urban population, and as discussed in Sect. 1.3, it is also necessary to consider population migration between cities. There is still not enough evidence whether the population size of large cities is excessive or not [11], so further research is needed. However, in the midst of the corona disaster, the risk of concentration in large cities is being reviewed, such as the headquarters function of companies moving out of the metropolitan area for the first time in 11 years [12], and there are also indications that the necessity of population concentration in large cities for production activities has disappeared due to IT technology [13]. Analysis of the factors of population migration between cities nationwide, and policy consideration based on the equilibrium state as a result, will be necessary. Cost-benefit analysis of public works has contributed to improving administrative accountability, but basically, regions with high demand have high priority. If public works in large cities are prioritized as a result of cost-benefit analysis, it will temporarily alleviate congestion in large cities and improve utility. However, if the utility curve for city size is assumed to be the shape of Fig. 1.4b, population migration may lower the productivity of small- and medium-sized cities and worsen congestion in large cities. As a result, that would potentially reduce the utility of the entire country. Even in the consideration of measures for cities and transportation, the perspective of population migration between cities is necessary, and for that, empirical evidence of the utility curve for the size of cities nationwide is indispensable.

References

1. Kii M, Akimoto K, Doi K (2012) Random-growth urban model with geographical fitness. Physica A Stat Mech Appl 391(23):5960–5970
2. Kii M (2021) Projecting future populations of urban agglomerations around the world and through the 21st century. npj Urban Sustain 1(1)
3. Kanemoto R (1997) Urban economics. Toyo Keizai Shinposha, Tokyo, p 377
4. Ministry of Land, Infrastructure, Transport and Tourism, Measures to Correct Overconcentration in Tokyo. https://www.mlit.go.jp/policy/shingikai/content/001374933.pdf
5. Kii M, Tamaki T, Suzuki T, Nonomura A (2023) Estimating urban spatial structure based on remote sensing data. Scientific Reports volume 13, Article number: 8804
6. Urban Railway Policy Division, Transition of Average Congestion Rate in Major Sections of the Three Major Metropolitan Areas, Railway Bureau (2021) Ministry of Land. Infrastructure, Transport and Tourism
7. Ministry of Internal Affairs and Communications, White Paper on Information and Communications. 2021
8. Industrial Labor Bureau, Results of a Survey on Telework. 2022, Tokyo Metropolitan Government
9. Choo S, Lee T, Mokhtarian PL (2007) Do transportation and communications tend to be substitutes, complements, or neither? Transport Res Record 2010(1):121–132
10. Tsukai M, Okumura M (1999) A gravity model of business travel and communication patterns considering substitutability and complementarity. J City Plan 34:85–90
11. Kanemoto R (2006. Autumn 2006 issue) Is Tokyo too large? Quarterly housing and land. Economics:12–20
12. Teikoku Databank, Survey on the trend of head office relocation in the metropolitan area. 2021
13. Kuma K (2021) 'Breaking away from the box'—architecture is the OS (Operating System) of society, in Monthly Keidanren. Keidanren, pp 52–60

Open Access This chapter is licensed under the terms of the Creative Commons Attribution-NonCommercial-NoDerivatives 4.0 International License (http://creativecommons.org/licenses/by-nc-nd/4.0/), which permits any noncommercial use, sharing, distribution and reproduction in any medium or format, as long as you give appropriate credit to the original author(s) and the source, provide a link to the Creative Commons license and indicate if you modified the licensed material. You do not have permission under this license to share adapted material derived from this chapter or parts of it.

The images or other third party material in this chapter are included in the chapter's Creative Commons license, unless indicated otherwise in a credit line to the material. If material is not included in the chapter's Creative Commons license and your intended use is not permitted by statutory regulation or exceeds the permitted use, you will need to obtain permission directly from the copyright holder.

Chapter 2
Land Use and Transportation

Akinori Morimoto

2.1 The Interrelationship of Land Use and Transportation

Understanding the interrelationship between land use and transportation is extremely important in designing a more prosperous society and safer transportation system [1].

The relationship between land use and transportation is similar to the relationship between a chicken and an egg, as they are interdependent upon each other (Fig. 2.1). Urban activities occur by utilizing land and transportation facilities such as roads that are developed to accommodate new transportation activities. Newly constructed or widened roads increase the attractiveness of the land along the road, inducing new urban facilities. When a city is growing slowly, it is easy for land use planning and transportation planning to keep pace. However, when rapid economic growth occurs, the demand for land use accelerates, while the development of transportation facilities cannot keep up. In a high-growth economy, many cities will experience severe traffic congestion and other transportation problems.

So, how can we balance land use and transportation? In our country, during times when the development of transportation facilities could not keep up with the road demand expanding due to rapid motorization, measures were implemented to adjust the transportation demand itself. This is called traffic demand management (TDM). When roads cannot be newly constructed or widened immediately, the peak demand was changed, or a shift to other means of transportation was attempted, making the most of existing road facilities.

A more fundamental approach lies in the proper guidance of land use. When transportation facilities are poor, the floor area ratio can be kept low, and the balance can be achieved by relaxing the floor area ratio according to the state of transportation infrastructure development. Also, changing the city structure itself for long-term sustainability should be considered. By dispersing the transportation demand

---

A. Morimoto (✉)
Faculty of Science and Engineering, Waseda University, Shinjuku, Japan
e-mail: akinori@waseda.jp

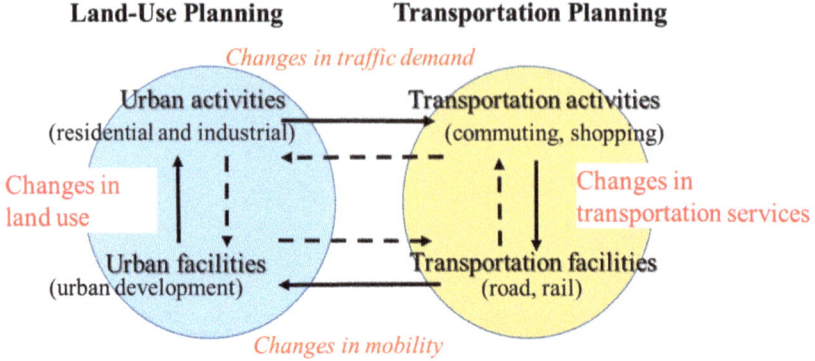

Fig. 2.1 Interrelationship between land use and transportation planning

concentrated in the city center to sub-centers and core cities, traffic congestion can be alleviated while aiming for balanced development of the entire city. This has been implemented in large cities, including Tokyo, as a polycentric dispersion type of urban structure. The Third National Capital Region Development Plan (1976), for example, recommends a more wide-area, multipolar distribution to correct Japan's overconcentration on Tokyo, and the follow-up Fourth Development Plan (1986) suggests prioritizing development of business core cities and secondary core cities with the goal of creating a more multi-zonal, multinuclear urban structure.

Additionally, transportation issues such as congestion, noise, and air pollution vary depending on the region and era. In the twenty-first century, the transportation issues that our country is facing include dealing with a super-aged society and global environmental issues, reflecting a stronger long-term and comprehensive perspective. However, even for such new problems, considering land use and transportation planning over time can lead to finding solutions to a variety of transportation problems so as to meet the needs of the future.

What kind of urban structure should be aimed for, and by what method should such an urban structure be developed? What kind of transportation will be emphasized at that time? How should we deal with the local congestion that still occurs? In this section, we will learn about the four keywords necessary for the next generation of urban development, based on the measures taken against the issues experienced by advanced countries.

2.2 Compact City

2.2.1 Why Is a Compact City Necessary?

How to organize and integrate expanding urban areas due to the progress of motorization into sustainable cities is a major issue in mature societies. Especially in Japan, which has entered a decline in population, it is crucial to smartly reduce unnecessary urban development in cities (Smart Shrink). The Second Report of the Social Capital Development Council (2007) points out the following problems if the diffused urban structure is left unattended:

- Difficulty in Maintaining Public Transportation: It is impossible to secure a substantial demand in low-density urban areas.
- Mobility Issues in a Super-Aged Society: The number of transportation-disadvantaged people who cannot use cars is increasing.
- Increasing Environmental Burden: Excessive dependence on cars increases environmental burden.
- Further Decline of the City Center: The relative attractiveness decreases due to the promotion of suburban development and sprawl.
- Pressure on Urban Finance: Increase in maintenance costs for diffused urban areas.

Even in developing countries that are experiencing economic growth, it is important to guide them to an appropriate urban structure during the period of population influx in order not to follow the same path as many advanced countries.

2.2.2 What Is a Compact City?

A compact city is a "sustainable urban structure that is friendly to people and the environment by consolidating the functions necessary for life in the center, maintaining an appropriate population density." In the population increase period, urban areas have expanded by encroaching on the surrounding green spaces, while in the population decrease period, it is expected that the city itself will shrink to an appropriate size by the green spaces encroaching on the urban areas (Fig. 2.2). In

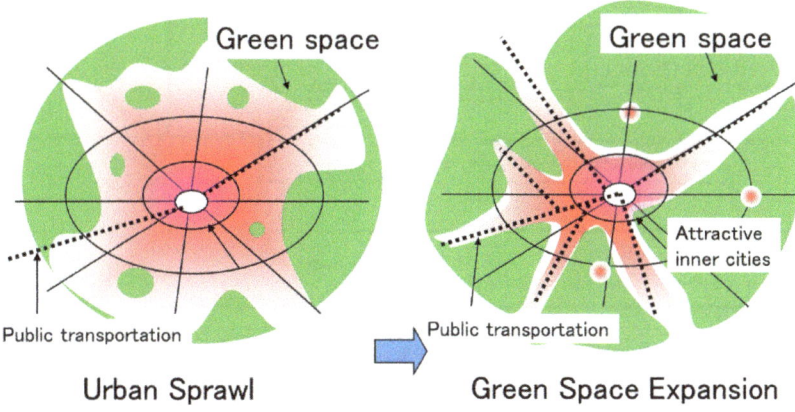

Fig. 2.2 Smart shrink

particular, consolidation along public transportation axes is important in considering the proper balance between cars and public transportation.

2.2.3 Proposal of Network-Type Compact City

The concept of a compact city gained attention following its recommendation as a model for sustainable development in the 1987 United Nations Brundtland Report. The aim was to create a city where people could live without overly relying on private cars, instead using public transportation and walking. However, it is extremely difficult to consolidate once expanded urban areas, and while each municipality sets this as a goal in their urban planning master plans, the road to realization is long. In order to break away from the low-density, car-oriented urban structure, further promotion of community development centered on public transportation is essential. Here, we propose a "network-type compact city" based on the concept of transit-oriented development (TOD) described later [2] (Fig. 2.3).

A network-type compact city is a city that "consolidates (compacts) the diverse attractions within the city at multiple hubs, and links them with various modes of transportation centered on convenient public transportation (networking)." The image of the urban structure is shown in Fig. 2.4. The term "compact" here does not necessarily mean concentration at a single point, but efficient consolidation at multiple hubs. Network-type compact cities are also resilient to disasters. By connecting the consolidated hubs, even if part of the city is affected by a disaster, the redundancy within the city can be ensured, and other areas can flexibly carry out recovery activities, which also enhances the resilience of the entire city.

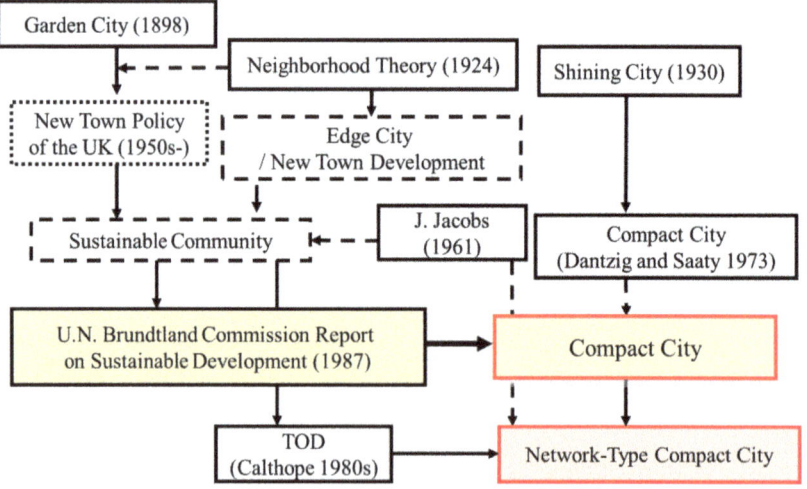

Fig. 2.3 A genealogy of network-type compact cities [10]

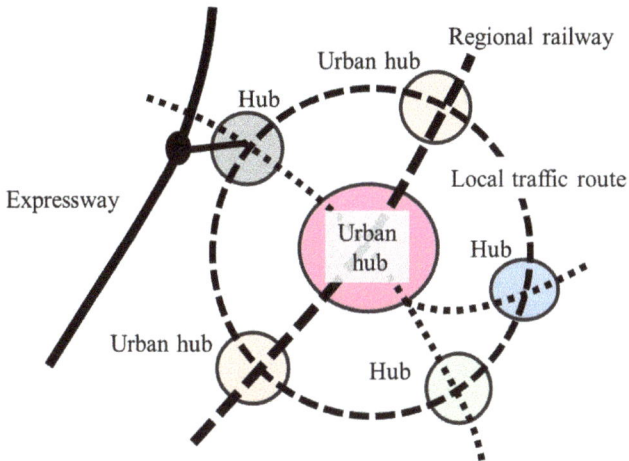

Fig. 2.4 A concept of network-type compact city

2.3 Transit-Oriented Development (TOD)

2.3.1 The Merits and Demerits of Car-Dependent Urban Development

In times of rapidly increasing urban population, cities were plagued by a shortage of residential areas, and large-scale new towns were successively developed in the suburbs. In the suburbs of large cities, development along railway lines was common and increased along with the expansion of public transportation networks. On the other hand, the situation in regional cities was quite different, with many new towns developed regardless of the convenience of public transportation, leading to a significant increase in car dependency. In commercial development as well, convenience for cars was prioritized, with locations successively established around suburban bypasses, accelerating the decline of downtown areas.

While cars are indeed a comfortable and convenient mode of transportation, excessive dependence on them drives out other modes of transportation, creating a society where it is difficult for those without cars, the transportation disadvantaged, to live. Learning from this, urban development centered on public transportation is attracting interest in advanced countries.

2.3.2 What Is Transit-Oriented Development?

Transit-oriented development (TOD) refers to "urban development that does not overly depend on cars and is based on public transportation." Its fundamental concepts were proposed in the 1980s by Calthorpe, but precursors to this concept can

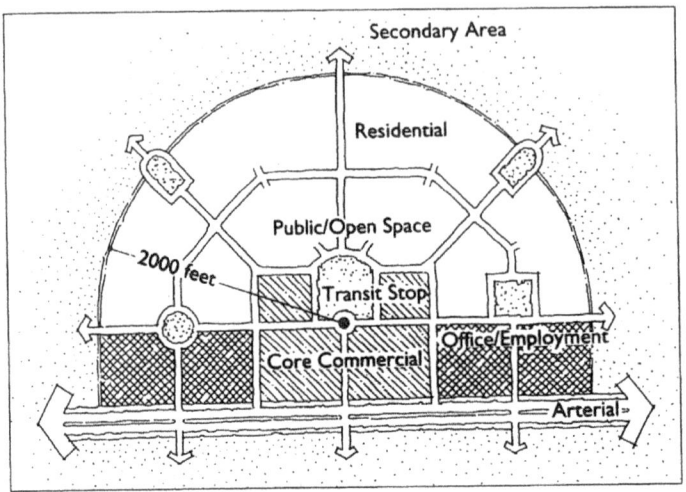

Fig. 2.5 Conceptual diagram of TOD (P. Calthorpe) [4]

be seen in Japan in the development of areas around train stations, promoted by private railway companies since before World War II. The aim is to create a town where people can live by walking, by placing functions such as commerce, business, and housing within a radius of about 600 m centered on the railway station (Fig. 2.5) [4].

There are three important elements (3Ds) in conducting transit-oriented development [5].

1. Density: A certain level of population density is necessary to maintain public transportation. The value varies depending on regional characteristics, but generally, it should be planned not to fall below 40 people/ha.
2. Diversity: It is important to place commercial functions and public functions such as medical and welfare services in front of the station and to consolidate basic functions of daily life within walking distance.
3. Design: Quality spatial design is indispensable for guiding land use. Attractive spaces change people's residential choice behavior (Fig. 2.6).

2.3.3 Aggregation of Urban Areas Using TOD

High-speed public transportation systems that connect with other cities, such as bullet train and railways, are established in the city center, and public transportation that combines punctuality and speed, such as LRT (light rail transit) and BRT (bus rapid transit), is introduced from the city center to the suburbs. TOD is mainly implemented on this public transportation axis [6]. This area is provided with high public transportation services on a regular basis for a long period of time, promoting

Fig. 2.6 San Francisco (Fruitvale Transit Village)

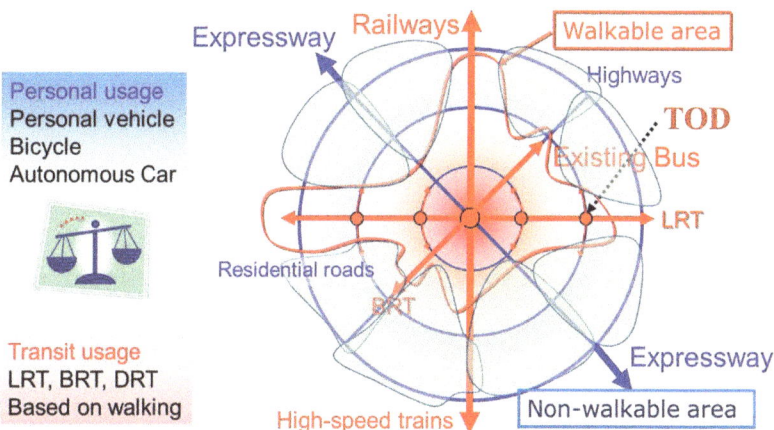

Fig. 2.7 Urban space design implementing TOD

urban aggregation. On the other hand, in the suburbs, variable public transportation services such as DRT (demand-responsive transit) are implemented in accordance with population decline (Fig. 2.7).

The TOD concept encourages the elderly and other transportation-disadvantaged people to live as much as possible within TOD, and families who enjoy the convenience of cars raise children in suburban green urban areas. In other words, by choosing a place to live according to changes in lifestyle, we aim to create a city that can accommodate all generations.

2.4 Next-Generation Transportation System

2.4.1 *Reconstructing the Hierarchy of Urban Transportation*

In regional cities, there are many cities where the only means of transportation inside and outside the city has become cars due to excessive dependence on cars. It can be said that the hierarchy of urban transportation has been significantly destroyed, such as through traffic that has penetrated into living roads and bus transportation that has been forced to withdraw. It is an urgent task to reconstruct a transportation system that does not rely solely on cars in order to respond to a super-aged society that is friendly to the environment.

The desirable transportation system for the next generation should balance automobiles and public transportation and should functionally differentiate in a hierarchical manner. In terms of automobile traffic, this hierarchy ranges from highways that maximize traffic function to local roads with high access function, forming a stepwise pyramid structure. On the other hand, public transportation also has a similar pyramid structure, from high-speed intercity Shinkansen and railways to community buses and shared taxis that move within districts. As the pyramid structure goes down, the speed of movement decreases, while the flexibility of the service increases (Fig. 2.8).

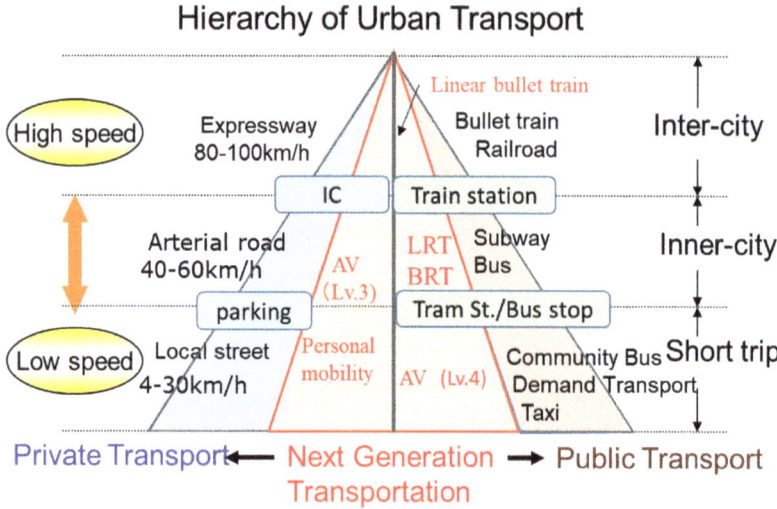

Fig. 2.8 Next generation transportation and urban transportation hierarchy

2.4.2 Next-Generation Public Transportation System

To revitalize declining urban public transportation, it is necessary to introduce a more comfortable and convenient public transportation system. This requires the following functions:

1. Punctuality: A system that has a dedicated lane and is not affected by traffic congestion.
2. Comfort: A system with minimal vibration and barrier-free, friendly to the elderly.
3. Sustainability: A system that can operate with less energy and produce less noise and emissions.
4. Attractiveness: A system where vehicles and their stops fit into the cityscape and contribute to city planning.

It is important that the system has not only the function to induce a switch from cars but also the added value to change the land use along the route.

The light rail transit (LRT) and the bus rapid transit (BRT) are attracting attention as next-generation public transportation systems with the above functions.

In contrast to heavy rail systems that provide transportation between cities, LRT systems provide transportation within a city. In Japan, these often take the form of next-generation tram systems. It refers to those that have improved the performance of traditional trams, strengthened cooperation with other means of transportation, and contributed to city planning as a comprehensive urban transportation system (Fig. 2.9). The LRT developed in Edmonton, Canada, in 1978 is considered the first. Since then, the number of cities introducing LRT has been increasing worldwide, reaching 111 cities as of 2008 [7].

BRT systems provide large-scale, rapid transport of passengers by bus. Unlike traditional bus routes, these systems use dedicated traffic lanes to allow frequent and punctual service, which can in some situations improve transportation capacity through the use of articulated vehicles and allow for smoother boarding at dedicated stops (Fig. 2.10). Ottawa (Canada), Curitiba (Brazil), and Bogotá (Colombia) are cited as precedents.

A commonality between both systems is that they are not simply introduced as a means of transportation but are developed as part of a sustainable mobility system suitable for next-generation cities. It is expected that attractive public transportation will promote the concentration of land use while supporting public transportation from the perspective of land use.

Fig. 2.9 LRT in Houston

Fig. 2.10 BRT in Curitiba

2.4.3 Next-Generation Transportation and Human-Centered Transportation Systems

Individual transportation, including private cars, continues to evolve. Electric vehicles (EVs) that run on motors using electricity as an energy source and fuel cell vehicles (FCVs) that generate electricity using hydrogen as fuel have been developed and put into practical use due to innovations in automobile technology. In addition, autonomous driving technology is evolving daily, and as of the revision of the Road Traffic Act in April 2023, fully autonomous driving (Level 4) on public roads is permitted under certain conditions in Japan.

In addition, small mobility aids are also becoming popular. For example, the use of personal mobility, a one-person small mobility aid, as well as micromobility, is gaining popularity in advanced nations. Micromobility, which is even smaller, lighter, and slower, but greatly assists in short-distance movement, is increasing mainly in urban areas (Fig. 2.11).

Also, with the advancement of information and communication technology (ICT), services that combine various modes of transportation have begun. MaaS (Mobility as a Service), which started in Helsinki, Finland, in 2016, has revolutionized the way transportation services are provided around the world. In the future, with further development of AI and ICT, a new transportation system will emerge where users can freely choose from a variety of transportation methods, including existing systems and next-generation transportation, and move seamlessly. Here, we call this a "human-centered transportation system" (Fig. 2.12).

Fig. 2.11 Small mobility aids

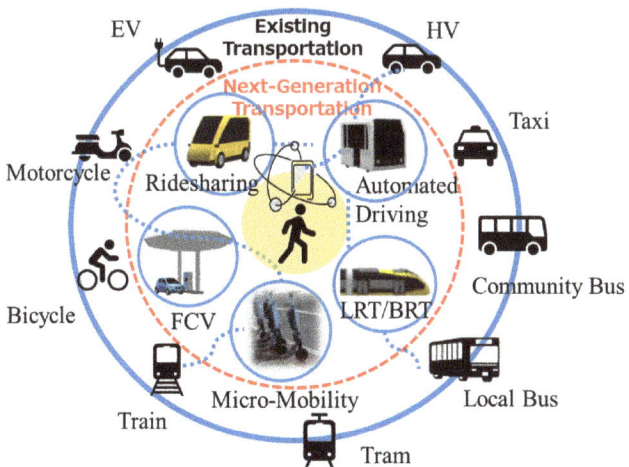

Fig. 2.12 Human-centered transportation system

2.5 Next-Generation Transportation Systems

2.5.1 Traffic Assessment and Simulation

From the perspective of harmonizing transportation and land use, the Traffic Impact Assessment (TIA) has been implemented since the 1990s as a method and system for pre-evaluating the traffic impact of development plans and taking necessary traffic measures [8].

In conducting a traffic assessment, it is extremely important to quantitatively and accurately understand the traffic situation. Recent innovations in information technology have led to a dramatic improvement in computer processing capabilities, making it possible to reproduce detailed traffic conditions on a second-by-second basis, incorporating signal control, lane configuration, etc. and to visually express them through animation. As a result, many traffic simulators have been developed both domestically (tiss-NET, AVENUE, TRAFFICSS, VISITOK, etc.) and overseas (NETSIM, Pramics, WATSim, TransModeler, etc.) and are being used in practical applications (Fig. 2.13).

Based on these results, by taking possible traffic measures among merchants, road managers, and traffic administrators, it is possible to mitigate the traffic impact before the establishment of large-scale stores. Note that there are many significant effects of large-scale development, and in situations where mitigation measures are difficult through transportation facility development, reconsideration of site selection for large-scale development may become necessary [9].

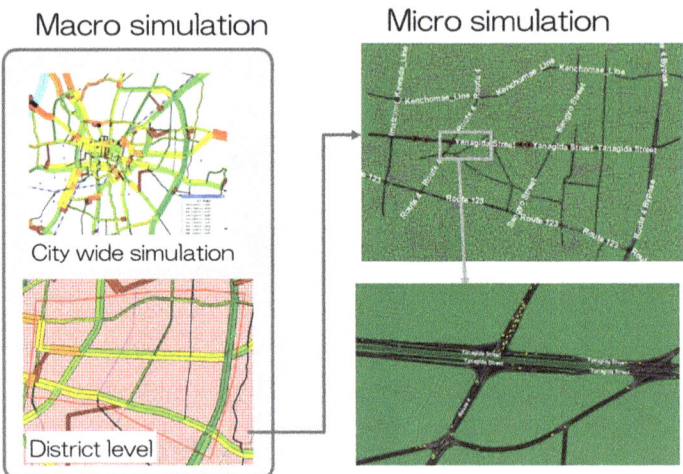

Fig. 2.13 Images of traffic simulation [10]

2.5.2 Smart City

In recent years, various initiatives using artificial intelligence (AI) and information and communication technology (ICT) have been attracting attention. Facilities and services that were previously owned and used individually can now be easily and efficiently shared by multiple people. As the sharing economy, such as car sharing and co-working evolves, the "smart city" utilizing ICT is also attracting attention in urban planning. A smart city is defined as "a sustainable city or district that optimizes the whole while utilizing new technologies such as ICT to address the various challenges faced by the city" (Fig. 2.14) [11].

Since the early 2010s, discussions on smart cities have been active, and there are many examples both in Japan and abroad. Initially, smart cities started with solving individual field problems, but in recent years, there has been an increase in broader, cross-field initiatives targeting the entire city.

2.5.3 Integration of Physical Space and Cyber Space

Discussions on urban planning, such as the concept of the compact city as a sustainable urban model in response to a declining population, is becoming more widely accepted. Furthermore, the smart city concept that aims for overall optimization using new technologies such as ICT is becoming more prevalent as well. Traditional urban planning, which was mainly about improving physical space (real space), has greatly expanded its scope and targets through the use of cyber space (virtual space). Therefore, in urban planning, physical space and cyber space should be considered simultaneously, as their integration is becoming an issue (Fig. 2.15).

For example, while a compact city aims to consolidate around a station, a smart city, which also aims to improve efficiency in areas away from the station, often raises concerns about potential trade-offs. If the appeal of the suburbs increases in a

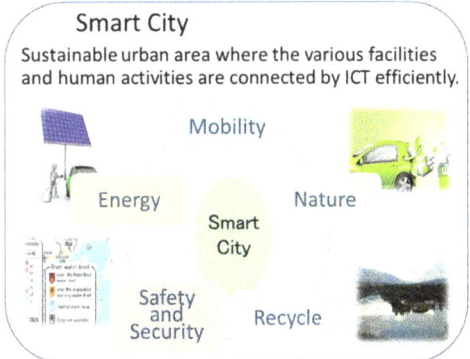

Fig. 2.14 Concept of smart city

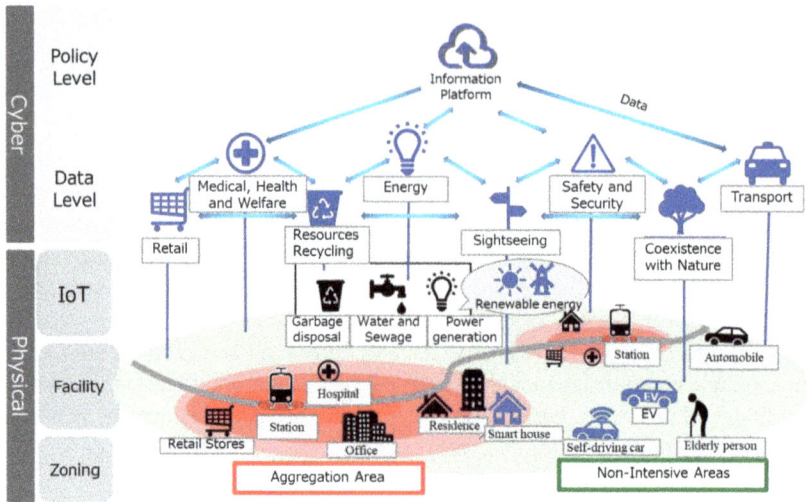

Fig. 2.15 Toward the integration of compact and smart cities [10]

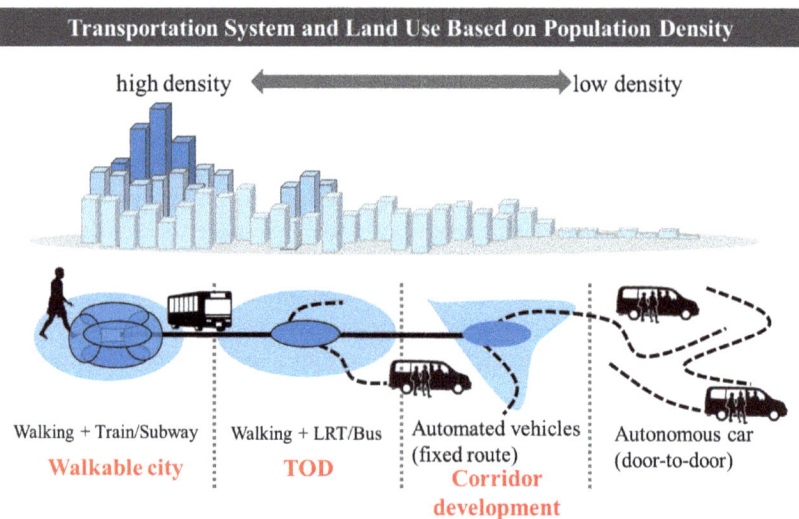

Fig. 2.16 Desirable transportation and land use relationships [10]

smart city, the relative appeal of the city center may decrease. The key is how to design and manage cyberspace to achieve overall optimization.

Various modes of transportation exist in cities, each with its own advantages and disadvantages. It is crucial to cross-integrate these effectively and provide them according to the characteristics of the city or district accordingly. The desirable relationship between transportation and land use is shown in Fig. 2.16. High-density city centers aim to create walkable spaces by focusing on walking and developing mass transit systems such as railways and subways. Urban areas carry out transit-oriented development (TOD), with light rail transit (LRT) and bus rapid transit (BRT) playing a vital role in urban trunk line transportation. Furthermore, the suburbs provide mobility services through scheduled route autonomous buses. The outer edges of the city provide door-to-door autonomous driving services, offering finely-tuned responses according to demand. Such a transportation system is expected to ensure the sustainability of the urban structure, maintain public activity, and increase people's well-being.

References

1. Tomorrow's Urban Transportation Policy (2003) Sugiyama et al (ed). Seibundo
2. Morimoto A (2012) A preliminary proposal for urban and transportation planning in response to the Great East Japan earthquake. IATSS Res 36(1):20–23
3. Transit Cooperative Research Program (2004) TCRP Report 102, transit-oriented development in the United States: experiences, challenges, and prospects, Transportation Research Board
4. Calthorpe P (1993) The next American Metropolis; ecology, community, and the American dream. Princeton Architectural Press
5. Developing Around Transit-Strategies and Solutions That Work, ULI (the Urban Land Institute). 2004
6. Hayashi R, Kato H, Doi K (2009) International association of traffic and safety sciences land use traffic research group edited 'quality stock of cities—integrated strategy of land use, green space, and traffic'. Kashima Publishing
7. Special Feature/Issues and Prospects for LRT (2009) Introduction to Japan. IATSS Review 34(2)
8. Research Report on Traffic Assessment. International Association of Traffic and Safety Sciences, 2001
9. Seki T, Morimoto A (2010) Organization and Future Prospects of Traffic Assessment in Large-Scale Development. Civil Engineering Society Paper Collection D 66(2):255–268
10. Morimoto A (2021) City and transportation planning: an integrated approach. Routledge
11. Ministry of Land, Infrastructure, Transport and Tourism, Urban Development Bureau, Towards the Realization of Smart City [Interim Summary], August 2018

Open Access This chapter is licensed under the terms of the Creative Commons Attribution-NonCommercial-NoDerivatives 4.0 International License (http://creativecommons.org/licenses/by-nc-nd/4.0/), which permits any noncommercial use, sharing, distribution and reproduction in any medium or format, as long as you give appropriate credit to the original author(s) and the source, provide a link to the Creative Commons license and indicate if you modified the licensed material. You do not have permission under this license to share adapted material derived from this chapter or parts of it.

The images or other third party material in this chapter are included in the chapter's Creative Commons license, unless indicated otherwise in a credit line to the material. If material is not included in the chapter's Creative Commons license and your intended use is not permitted by statutory regulation or exceeds the permitted use, you will need to obtain permission directly from the copyright holder.

Chapter 3
Transportation and the Environment

Tomohiro Ichinose

3.1 Impact of Global Environmental Issues and Pandemics

The issue of transportation and the environment began with the advent of the Industrial Revolution, when vehicles powered by fossil fuels were introduced. At that time, the impact of transportation equipment and infrastructure on the environment was limited. However, the rapid growth of motorization in the first half of the twentieth century led to the emergence of significant environmental problems in urban areas. These issues, which were initially confined to developed countries, became more pronounced after World War II. Air pollution caused by automobile exhaust gases led to the implementation of pollutant emission regulations in various countries and regions. However, with the expansion of the global economy, the utilization of natural resources, including fossil fuels, has increased exponentially, accompanied by a parallel surge in waste emissions. In 1972, the Club of Rome published a report titled "Limits to Growth," [1] which warned that if population growth and environmental pollution continue unabated, growth on Earth will reach its limit within 100 years.

The twenty-first century has been designated as the "century of the environment," yet global environmental issues remain a significant concern. The Planetary Boundary, developed by the Stockholm Resilience Center and Australian universities in 2009, identifies whether we have exceeded the limit in nine key areas: climate change, novel entities, stratospheric ozone depletion, atmospheric aerosol loading, ocean acidification, biogeochemical flows, freshwater use, land-system change, and biosphere integrity [2]. As of 2009, climate change, biogeochemical flows, and biosphere integrity had exceeded the limit (although two items could not be quantified at that time) [3]. By 2015, it was evident that land-system change had also exceeded the limit. The most recent data from 2023 indicates that novel entities and

---

T. Ichinose (✉)
Faculty of Environment and Information Studies, Keio University, Fujisawa, Japan
e-mail: tomohiro@keio.jp

© The Author(s) 2026
Pioneering the Future for Traffic and Safety Sciences,
https://doi.org/10.1007/978-981-96-0676-4_3

freshwater change have been added, thereby revealing that six out of the nine items have exceeded the limit [4]. The utilization of natural resources without limit is already beginning to exceed their limits. The limits of the Earth, previously warned by the Club of Rome, have become a reality. Climate change due to greenhouse gases (GHG), already considered to have exceeded the limit as of 2009, has become widely recognized as a global environmental issue that cannot be delayed. In August 2021, the Intergovernmental Panel on Climate Change (IPCC) released the first part of its sixth assessment report, IPCC [5]. This report demonstrated that from 2011 to 2020, the world's surface temperature rose by 1.09 °C compared to the period from 1850 to 1900. Furthermore, the report indicated that the majority of this temperature increase was due to human factors. The summer of 2023 witnessed record-breaking heat in numerous regions of the Northern Hemisphere. The World Meteorological Organization (WMO) and the European Union's meteorological information agency, the Copernicus Climate Change Service, have reported that July 2023 was the hottest month worldwide. The 26th Conference of the Parties (COP26) to the United Nations Framework Convention on Climate Change (UNFCCC) was held in Glasgow, UK, in November 2021. At this meeting, it was agreed that efforts should be made to limit temperature increases to 1.5 °C above preindustrial levels. Nevertheless, a report [6] released by the WMO in May 2023 indicates that there is a 66% probability that the 1.5 °C limit will be exceeded by 2027.

Climate change has a significant impact on transportation, primarily through vehicles, and transportation infrastructure has a profound impact on the natural environment. This impact has been linked to the biosphere integrity, which was already considered to have exceeded the Earth's limits in 2009, and land-system change, which was considered to have exceeded the limits in 2015. The 15th Conference of the Parties (COP15) to the Convention on Biological Diversity (CBD) was held in Montreal, Canada, in December 2022. At this meeting, a new target for 2030 (the Kunming-Montreal Global Biodiversity Framework) was agreed upon. The previous targets were the Aichi Targets, which consisted of 20 individual targets agreed upon in Nagoya in 2010. It has been revealed that the majority of these targets were not achieved by 2020 [7]. One of the new targets is called 30by30, which aims to increase the proportion of protected areas that contribute to biodiversity conservation to 30% by 2030. This target applies not only to land but also to marine areas.

The coronavirus disease 2019 (COVID-19), which has been a global phenomenon since the end of 2019, has had a significant impact on transportation. By the end of 2023, when this book was written, restrictions due to the pandemic had been lifted in almost all countries and regions. However, there are still reports of increasing numbers of cases of variant infections. Consequently, it can be argued that the effects of the pandemic are still being felt. Even prior to the pandemic, there existed a concept known as One Health, which posits that the health of humans, animals, and the environment must be safeguarded collectively [8]. In the wake of the global spread of the novel coronavirus, the manner in which transportation and urbanization are evolving across various regions is undergoing a profound transformation.

This phenomenon presents an opportunity to integrate environmental considerations into the planning and implementation of transportation and urban development strategies.

3.2 Past Transportation and Environmental Issues

In 1988, the OECD published a report titled "Transport and the Environment." [9] This was also translated into Japanese and published in 1993 [10]. The report presents a comprehensive analysis of the relationship between transportation and the environment. Table 3.1 is a revised version by the author based on the table published in the report, which organizes the impact of transportation on the environment.

The impact of transportation on the environment can be divided into two main categories: the impact of transportation infrastructure, such as roads, and the impact of vehicles and other means of transportation. With regard to the atmosphere, the impact of vehicles is the most significant, but the impact on land is due to the development of transportation infrastructure. This can be divided into two subcategories: the construction of facilities at hubs such as ports, stations, and airports, and the construction of linear routes such as canals, railways, and roads. The necessity for continuous routes inevitably results in the division of the natural environment and land use. Furthermore, the construction of large-scale facilities such as airports can have a significant impact on the natural environment and land use.

With regard to the impact on the atmosphere, it is evident that all modes of transportation are associated with carbon dioxide emissions and air pollution. During the latter half of the twentieth century, while industrial pollution, primarily sulfur oxides, began to improve, urban and lifestyle air pollution caused by sources such as automobiles became apparent. In 1962, the world's inaugural automobile exhaust regulations were implemented in California, USA, and countries around the globe, including Japan, have since enacted more stringent regulations, reducing emissions of carbon monoxide, hydrocarbons, nitrogen oxides, lead compounds, and particulate matter (such as PM2.5). Consequently, urban and lifestyle air pollution has been significantly improved in developed countries, yet it persists as a significant concern in developing countries. At the time of its publication in 1988, this report noted that the issue of climate change and global warming due to GHGs was not yet a significant concern.

The impact of transportation infrastructure on water bodies is also significant. The deterioration of water quality in rivers and coastal areas has been well documented, and the construction of new infrastructure often results in a decline in water quality. Additionally, the construction of underground transportation systems, such as tunnels, can lead to groundwater pollution due to the division of groundwater veins and an increase in maintenance costs for drainage. There are numerous instances of water pollution caused by transportation.

The abolition and renewal of transportation infrastructure generates a considerable amount of waste, and each means of transportation must also be treated as

Table 3.1 The effects of transportation on the environment

Transportation mode	Air	Land	Water (including groundwater)	Waste	Noise, vibration	Ecological impact	Landscape
Maritime transportation and inland waterways	CO_2 emissions Air pollution	Port facility construction Canal construction	Port facility construction Drilling and dredging rivers and coasts	Facilities, ship disposal	Noise around the port	Expansion of invasive species by ballast water Endocrine disruption of organisms by paint on ships	Loss of natural coast, rivers Landscapes with ships
Rail transportation	Carbon dioxide emissions Air pollution	Railway and station construction	Division of underground water veins by tunnel construction, etc.	Facilities and vehicle disposal	Noise and vibration near stations and along railways	Habitat fragmentation Collision accidents	Disruption of natural and traditional landscapes Creation of new landscape resources (vehicles, bridges)
Road transportation	Carbon dioxide emissions Air pollution (in particular, fuel additives such as CO, HC, NOx, dust, and lead)	Construction of related facilities, including roads Terrain modification due to road construction Procurement of materials for road construction	Division of underground water veins by tunnels, etc. Development of water regions and changes in water system due to road construction Surface and groundwater contamination	Facilities and vehicle disposal, waste oil Battery disposal (especially for hybrid and electric vehicles)	Automobile noise and vibrations in cities and along major routes	Habitat fragmentation collision accidents Disruption due to pollutants Contamination by anti-freeze agents Light pollution by streetlights	Disruption of natural and traditional landscapes Creation of new landscape resources (bridges, etc.)
Air transportation	Carbon dioxide emissions Air pollution	Airport facility construction	Development of water regions due to airport construction	Disposal of aircraft	Noise and shock waves near airports	Habitat destruction by airport development Collisions with airplanes (mainly birds)	Loss of natural landscapes Landscapes with airplane

Adapted from Ref. [9], with additions by the author

waste when it exceeds its service life. In Japan, the Act on Recycling of End-of-Life Automobiles was enacted in 2003, making automobile recycling mandatory. The recent recycling rate is as high as 96.1% [11]. Nevertheless, with the accelerated proliferation of hybrid vehicles and the prospective advent of electric and next-generation environmentally friendly automobiles, the disposal of batteries is anticipated to surge exponentially, potentially leading to a concomitant surge in the cost of recycling.

Noise has been extensively studied in the context of transportation, with the exception of ships. Various measures have been attempted to mitigate its impact. When developing transportation infrastructure, it becomes subject to the Environmental Impact Assessment Act according to its scale and must meet the standards set by law. For each mode of transport, development continues to reduce noise and vibration.

The impact on living organisms has been discussed less in the transportation sector, and various points have been made in biology and ecology. The most significant impact is the fragmentation and reduction of habitats. As will be explained later, measures such as environmental mitigation and the ecology road (eco-road) have been taken as countermeasures. Other various impacts can be seen, but in maritime and water transportation, the characteristics are expanding the distribution area of alien species by ballast water and endocrine disruption (so-called environmental hormones) by the paint used on ships. Also, in air transport, collisions between airplanes and birds have become a significant problem, posing a risk to human life.

Finally, the impact on the landscape must be considered. This point is not mentioned at all in the OECD Report [9]. Not only the impact on natural landscapes but also the transportation infrastructure can have a significant impact on traditional and historical landscapes that have been cultivated until now [12]. On the other hand, a significant feature of the landscape is that the vehicles and infrastructure can become landscape resources. Even those who are not particularly interested in railways will likely find a photograph of a train approaching a bridge spanning a valley aesthetically pleasing. There is a plethora of examples where the transportation infrastructure itself, such as the Yokohama Bay Bridge or the Akashi Kaikyo Bridge, has become renowned for its scenic value.

3.3 Transportation and Climate Change

3.3.1 Carbon Dioxide Emissions from the Transportation Sector

The 21st Conference of the Parties (COP21) to the UNFCCC was held in Paris at the end of 2015, during which the Paris Agreement was agreed upon. The Paris Agreement represents the first international framework for climate change in which all countries participate. A goal of limiting the average temperature rise to 2 °C by

2100 has been set as a common long-term goal for the world, with an effort goal of 1.5 °C. Each country has submitted its target reduction amount for how to reduce GHG emissions.

Carbon dioxide, a significant contributor to GHG emissions from human activities, is expected to reach 36.8 billion tons worldwide by 2022. In 2020, the global impact of the COVID-19 pandemic resulted in the restriction of various activities worldwide, leading to a reduction in carbon dioxide emissions from 36.2 billion tons in the previous year to 34.3 billion tons. However, with the lifting of travel restrictions in many countries, emissions increased to 36.5 billion tons in 20211 [3]. The total carbon dioxide emissions in the world in 1900 were estimated to be 2 billion tons. Although there were years when it temporarily decreased, it has consistently shown a rapid increase. Of this, the transportation sector accounted for 21.7%, following the electricity and heat sector (39.8%) and the industrial sector (24.9%) in terms of large proportions [13].

In Japan, the total GHG emissions are calculated in terms of carbon dioxide equivalent each year. In recent years, there has been a consistent decrease from the peak of 1.409 billion tons in fiscal 2013. Fiscal 2019 was 1.21 billion tons, and the impact of the corona disaster strongly reduced it to 1.147 billion tons in fiscal 2020. However, as restrictions were gradually lifted in fiscal 2021, emissions increased to 1.170 billion tons. Of this, the amount of carbon dioxide emissions was 1.064 billion tons. Figure 3.1 illustrates the ratio of emissions by each sector. The transportation sector accounted for 16.7% of emissions, followed by the electricity and heat sector (40.4%) and the industrial sector (25.3%), which is similar to the trend observed worldwide. When the carbon dioxide emissions from energy sources due to power generation and heat generation are converted into emissions allocated to each sector on the consumer side according to the consumption of electricity and heat, the ratio of the transportation sector increases to 17.4%. Furthermore, the carbon dioxide emissions from the transportation sector have exhibited a consistent downward trend in recent years. These emissions have decreased from their peak in fiscal 2001 to fiscal 2020 and have increased by 0.8% compared to the previous fiscal year in fiscal 2021.

The carbon dioxide emissions from the transportation sector are calculated by summing the carbon dioxide emissions per unit of power generation for those using electricity and the carbon dioxide generated by burning fuel to drive transportation equipment. Therefore, the recent decrease in emissions is said to be influenced by the improvement in the fuel efficiency of automobiles. In contrast to the relatively stable consumption of diesel oil, which has remained largely unchanged, with the exception of fiscal 2020 due to the impact of the global pandemic, the trend in gasoline consumption has been consistently declining year on year. These carbon dioxide emissions from the transportation sector represent a significant proportion of the sector's overall emissions; however, it is important to note that emissions related to the manufacture of transportation equipment and transportation infrastructure development are not included in this analysis. Instead, these emissions are incorporated into the industrial sector. Additionally, the majority of automobiles and

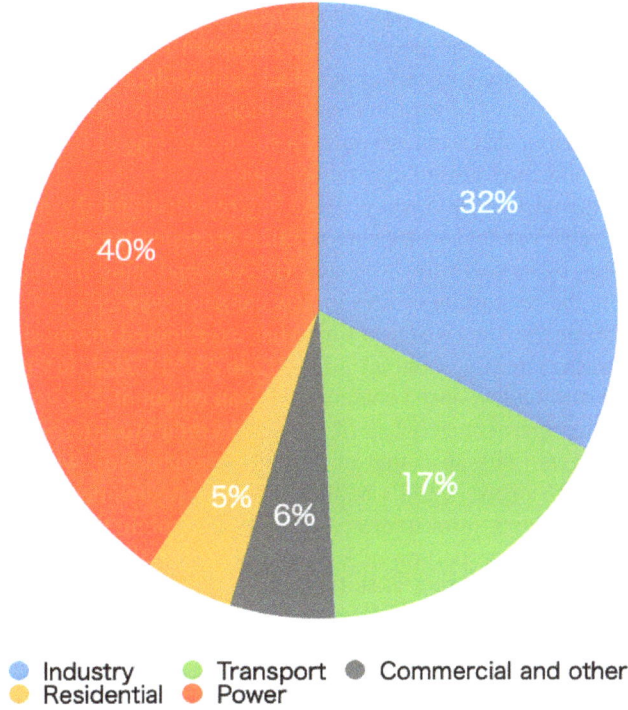

Fig. 3.1 Percentage of Japan's carbon dioxide emissions by sector in 2021 by https://www.env.go.jp/content/000128749.pdf

motorcycles sold by Japanese companies in Japan are manufactured abroad, resulting in the omission of emissions generated during the manufacturing process from the emissions data in Japan.

3.3.2 Carbon Neutral of Transportation Equipment

3.3.2.1 Popularization of Electric Vehicles

Since the conclusion of the Paris Agreement in 2015, active measures have been considered and implemented for carbon neutrality in the transportation sector, which has a large emission ratio. On July 6, 2017, the French government announced that it would stop selling gasoline and diesel passenger cars by 2040, a policy that has attracted significant attention worldwide. This policy is positioned in the climate plan announced in July 2017. Subsequently, analogous policies were announced by the UK and China. In July 2021, the European Commission proposed a regulation to reduce carbon dioxide emissions from new passenger cars and small commercial vehicles to zero by 2035. In October 2022, a policy to ban the sale of gasoline cars, including hybrid cars, and transition to electric vehicles (EVs) and fuel cell vehicles (FCVs) was passed in the European Parliament. In 2015, it was discovered that automobile manufacturers had been violating diesel vehicle emission regulations. This loss of trust in diesel vehicles, which had been a mainstream option in Europe until that point, served as a significant trigger for European countries to shift significantly toward electric vehicles.

The International Energy Agency (IEA) has reported that the number of new electric vehicles (EVs) sold worldwide in 2022 exceeded ten million for the first time, including battery electric vehicles (BEVs) and plug-in hybrid vehicles (PHEVs) [13]. The proportion of EVs in new car sales was 29% in China, 21% in Europe, and 8% in the United States. China has experienced the most rapid spread of EVs, with sales increasing from 0.1% in 2012 to 29% in 2022. In China, there are more than one million low-speed chargers (with an output of 22 kilowatts or less), which account for approximately half of the world's total. Additionally, there are 76 high-speed chargers (with an output of 22 kilowatts or more), which account for approximately 90% of the world's total [13]. These figures demonstrate that the development of charging infrastructure is rapidly promoting the spread of electric vehicles. In Japan, the proportion of BEVs in new car sales in 2022 was 1.72%, while that of PHEVs was 1.10% [14]. Even when combined, the proportion is less than 3%, indicating that the spread is slower than in China and the West.

BEVs run solely on electric energy and do not emit carbon dioxide. Because BEVs can be powered by the existing power generation and transmission infrastructure, they are a strong option for achieving carbon neutrality soon. However, it has long been pointed out that the production of batteries and motors requires substances and metals known as rare earths and rare metals, which can cause environmental problems different from climate change. Moreover, the achievement of carbon neutrality at the national or regional level is contingent upon the source of electricity supplied. For instance, France, the first country to announce the cessation of gasoline and diesel vehicle sales, generated 69% of its power in 2021 through nuclear energy. The next largest proportion, 12%, was attributed to hydroelectric power, while power generation utilizing fossil fuels emitting carbon dioxide accounted for only 7.4%. In fiscal 2019, Japan's power generation was 37.1% natural gas, 31.9% coal, and 6.8% oil, among other sources, with 75.8% derived from fossil energy. Holland et al. [15] have investigated the environmental benefits of BEV introduction in the United States and have revealed that the environmental benefits vary significantly by region. In California, where there is less power generation by fossil energy, these benefits are maximized. Nevertheless, in regions such as North Dakota, where the majority of power generation is derived from coal, BEVs may potentially cause more significant environmental harm than gasoline vehicles. According to Japan's power source composition for fiscal 2022, 27.8% is still covered by coal, and the rapid introduction of BEVs at this point could be counterproductive.

3.3.2.2 Utilization of Hydrogen

Hydrogen is a fuel that is attracting attention in the pursuit of achieving carbon neutrality. With regard to automobiles, fuel cell vehicles (FCVs) have already been commercialized and are being sold. FCVs generate electrical energy through a chemical reaction between hydrogen and oxygen. In the sense of using electrical power, they fall under the category of EVs. FCVs do not emit not only carbon dioxide but also other air pollutants. Hydrogen cars that burn hydrogen itself to gain propulsion already exist, but they are generally not commercially available. Hydrogen is a fuel that does not emit any carbon dioxide when burned, but it has a high risk of explosion and requires careful handling. Furthermore, unlike oil, it is not a fuel obtained by mining but is industrially produced using natural gas. Consequently, the significant challenges associated with the cost of production, transportation, and storage, as well as the emission of carbon dioxide during the production process, must be addressed. As of January 2023, only 163 hydrogen stations in Japan supply hydrogen to FCVs. The majority of these stations are concentrated in major metropolitan areas. Although the number of gasoline stations is decreasing annually, there are still approximately 29,000 nationwide. While the number of charging stations is a challenge for BEVs, they can be easily charged at home if the time required is acceptable. The hydrogen supply system is a critical factor in the widespread use of FCVs and hydrogen cars. Nevertheless, the Japanese government has indicated that hydrogen power generation is essential for achieving carbon neutrality by 2050, which may result in further development of the hydrogen supply system.

3.3.2.3 Utilization of Other Alternative Fuels

A variety of fuels have been utilized for vehicles, including automobiles. In light of the growing demand for carbon neutrality, fuels other than electricity and hydrogen are being considered. In terms of utilizing plant-derived biomass, ethanol and biodiesel fuels have already been commercialized. However, there were once charcoal cars that burned biomass directly. Nuclear power has already been commercialized for ships, and the use of wind power is being reconsidered for ships.

Synthetic fuels have recently attracted significant attention due to the demand for carbon neutrality as soon as possible. Synthetic fuels synthesized from carbon dioxide and hydrogen can utilize carbon dioxide emitted using fossil fuels. Direct Air Capture (DAC) technology may directly separate and recover atmospheric carbon dioxide. It is believed that carbon recycling can be achieved if synthetic fuels can be produced using DAC. Synthetic fuels can be used by engines and other conventional internal combustion engines, making them an easy introduction. The most significant challenge is cost, which is considerably higher than that of existing fossil fuels. However, cost reductions are believed to be possible with the development of manufacturing technology, and numerous projects are currently being implemented, primarily in Europe and the United States. In this context, the EU, which had previously planned to prohibit the sale of all new cars powered by fossil fuel engines by 2035, revised its policy in March 2023 to permit the sale of engine cars utilizing synthetic fuels. The fact that the policy was changed in less than a year suggests that various factors are at work. However, it could also be said to show that carbon neutrality in the transportation sector cannot be achieved by one measure alone.

3.3.3 The Way of Transportation and Cities to Achieve Carbon Neutrality

In order to achieve carbon neutrality in the transportation sector, it is essential to reduce the amount of carbon dioxide emissions per transport vehicle. However, if the total amount increases, reducing greenhouse gas (GHG) emissions will not be possible. Figure 3.2 illustrates the trend in the number of new passenger car registrations in different world regions over the past decade. There has been little change in Europe and the Americas, and the number has decreased in recent years. However, in Asia and Oceania, the trend is increasing. Although it declined due to the impact of the COVID-19 pandemic, it is growing again. Africa is difficult to gauge due to its smaller total number compared to other regions. Nevertheless, in 2022, the number of cars owned in Africa was 1.8 times higher than in 2011. While the number of cars owned in developed countries has plateaued, new registrations have surged in recent years in China and India.

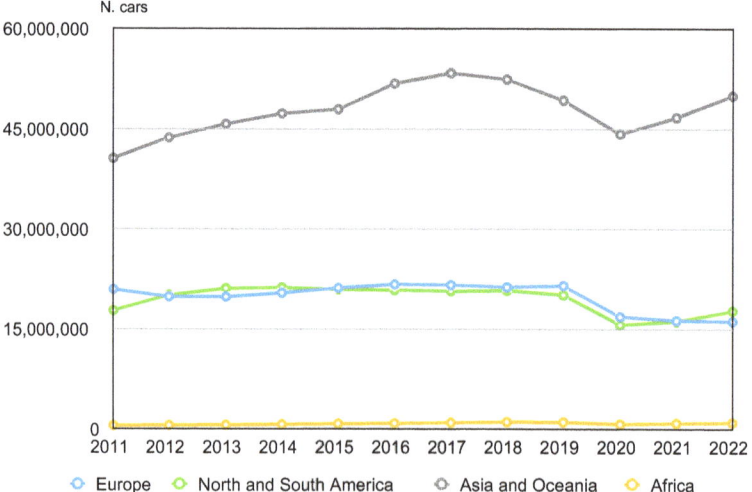

Fig. 3.2 New car registrations by region of the world. (The author made from data of http://www.oica.net/)

Developed countries, led by the United States, have fundamentally altered their urban planning strategies in response to the advent of the automobile. The grid-based urban design that emerged during this period assumes the dominance of the automobile and is incompatible with the movement of people on foot, by bicycle, or via public transportation. Since the 1990s, however, there has been a shift toward cities that prioritize non-motorized transportation. The goals of this transition are diverse, including the reduction of air pollutants, the alleviation of traffic congestion, the promotion of health, the revitalization of downtown areas, and the formation of communities. In particular, in cities that experienced lockdowns during the COVID-19 pandemic, restrictions on car travel led to a reevaluation of walking and cycling. The 15-minute city concept proposed by Moreno et al. [16] was proven to be of high interest to citizens when Anne Hidalgo, who included the 15-minute city in her campaign promises, was elected mayor of Paris in 2020. The transition from automobile use to walking, cycling, and public transportation, known as the modal shift, has been demonstrated by numerous studies to contribute to the reduction of air pollutants and congestion, and to result in a decrease in carbon dioxide [17]. In countries such as China, where urban areas have been experiencing rapid expansion in recent years, investments have been made in public transportation, including subways. Nevertheless, it has been observed that the advantages of these investments are more pronounced in alleviating traffic congestion than in reducing air pollution [18].

Mobility as a service (MaaS), which has been increasingly introduced in Western countries in recent years, also plays a significant role in achieving carbon neutrality. MaaS is a service that enhances the convenience of users' mobility by connecting all services related to transportation, including not only public transportation but also bicycles, micro-mobility, and car sharing. It is also making a significant contribution to reducing carbon dioxide emissions. The European Union (EU) has set a goal of formulating an urban transportation strategy by 2025 that includes new mobility services. This strategy is intended to achieve a 55% reduction in greenhouse gas (GHG) emissions by 2030. The strategy will rely on the use of MaaS, with smartphone apps serving as the primary platform for implementation. The apps will display the amount of carbon dioxide emissions in real time, which is expected to encourage users to alter their behavior. The app provides data and serves as a tool for collecting data from users, including their movements, which enables the provision of data-driven services. This will also enable real-time data collection on environmental impact, with greater accuracy than ever. While various discussions and caution are required regarding the handling of personal information such as individual travel routes and services used, it has become possible to envision a transportation vision that provides mobility services based on accurate data and aims for carbon neutrality at the same time.

3.4 Impact of Transportation Infrastructure on the Natural Environment

3.4.1 Environmental Impact Assessment

As previously stated in Sect. 3.2, the development of transportation infrastructure is accompanied by changes in land use, which has a multitude of effects on the natural environment. The rapid increase in vehicles since the conclusion of World War II has been supported by the rapid development of transportation infrastructure. Such infrastructure development has naturally had a significant impact on the natural environment. In the United States, the National Environmental Policy Act (NEPA) was enacted in 1969 with the objective of institutionalizing environmental impact assessments (EIA) in order to minimize the impact of various developments on the natural environment. Japan, however, lagged behind Western countries in this regard, with the Environmental Impact Assessment Law enacted in 1997. Table 3.2 lists projects subject to environmental assessment in Japan. The businesses targeted are divided into 13 categories, including port planning, which has 14 categories. Among them, roads, railways, airports, and port planning are indeed transportation infrastructure. In addition, land readjustment projects, new housing urban development projects, industrial park development projects, new city infrastructure development projects, distribution business park development projects, and residential land development projects are also categories that greatly involve transportation infrastructure as the foundation of the city. As a result, it is easy to understand the magnitude of the impact of transportation infrastructure on the environment. However, not all projects are subject to environmental assessment. As shown in Table 3.2, projects of a particular scale or larger are targeted. For Type 2 projects, the necessity of an environmental assessment is determined on a case-by-case basis. Additionally, numerous local public entities have enacted separate environmental impact assessment ordinances, and projects not subject to the Environmental Impact Assessment Law are also subject to assessment under the ordinance.

In environmental assessments, various items are investigated, predicted, and evaluated. Taking the Kanagawa Prefecture Environmental Impact Assessment Ordinance as an example, it targets the following 21 items. That is, air pollution, water pollution, soil pollution, noise and low-frequency noise, vibration, ground subsidence, odor, waste and generated soil, radio wave interference, sunlight obstruction, reflected light (limited to sunlight reflected by solar panels), weather, water phenomena, earth phenomena, plants, animals, ecosystems, cultural properties, landscapes, recreational resources, greenhouse gases, regional division, and safety. Some items, such as reflected light, are only involved in specific development projects, but almost all evaluation items will be targeted when developing transportation infrastructure.

In the case of projects subject to the Environmental Impact Assessment Law, the environmental assessment procedure will go through five major stages. The creation of a consideration document that discusses and summarizes matters that need to be

3 Transportation and the Environment

Table 3.2 List of projects subject to the Environmental Impact Assessment Act [19]

		Class-1 project (EIA is always required)	Class-2 project (the necessity of EIA is judged by project)
1.	Road		
	National expressway	All	–
	Metropolitan expressway	4 lanes or more	–
	National roads	4 lanes or more, 10 km or longer	4 lanes or more, 7.5 km or more and less than 10 km
	Large-scale forest road	Width: 6.5 m or wider, 20 km or longer	Width: 6.5 m or wider, 15 km or more and less than 20 km
2.	River		
	Dam, weir	Reservoir area:100 ha or larger	Reservoir area: 75 ha or more and less than 100 ha
	Diversion channel, lake-related development	Area of land alteration:100 ha or larger	Area of land alteration: 75 ha or more and less than 100 ha
3.	Railway		
	Shinkansen (super express train)	All	–
	Railway, track	Length:10 km or longer	Length: 7.5 km or more and less than 10km
4.	Airport	Runway:2,500 m or longer	Runway: 1,875 m or more and less than 2500 m
5.	Power plant		
	Hydraulic power plant	Output:30,000 kW or over	Output: 22,500 kW or more and less than 30,000 kW
	Thermal power plant	Output:150,000 kW or over	Output: 112,500 kW or more and less than 150,000 kW
	Geothermal power plant	Output:10,000 kW or over	Output: 7500 kW or more and less than 10,000 kW
	Nuclear power plant	All	–
	Solar battery power plant	Output:40,000 kW or over	Output: 30,000 kW or more and less than 40,000 kW
	Wind power plant	Output:50,000 kW or over	Output: 37,500 kW or more and less than 50,000 kW
6.	Waste disposal site	Area:30 ha or larger	Area: 25ha or more and less than 30 ha
7.	Landfill and reclamation	Area: exceeding 50 ha	Area: 40 ha or more and 50 ha or less
8.	Land readjustment project	Area: 100 ha or larger	Area: 75 ha or more and less than 100 ha
9.	New Residential area development project	Area: 100 ha or larger	Area: 75 ha or more and less than 100 ha

(continued)

Table 3.2 (continued)

		Class-1 project (EIA is always required)	Class-2 project (the necessity of EIA is judged by project)
10.	Industrial estate development project	Area: 100 ha or larger	Area: 75 ha or more and less than 100 ha
11.	New town infrastructure development project	Area: 100 ha or larger	Area: 75 ha or more and less than 100 ha
12.	Distribution center complex development project	Area: 100 ha or larger	Area: 75 ha or more and less than 100 ha
13.	Residential or industrial land development by specific organizations	Area: 100 ha or larger	Area: 75 ha or more and less than 100 ha
	Port and harbor planning	Total reclaimed and excavated land: 300 ha or larger	

considered at the planning stage, the creation of a method document that outlines the proposed items and methods for conducting an environmental assessment, the creation of a preparation document that summarizes the proposed assessment results, the evaluation document that finalizes the assessment results by adding modifications to the preparation document, and the creation of a report that summarizes the results of environmental conservation measures taken when implementing the project. These consideration documents, method documents, preparation documents, evaluation documents, and reports are displayed for a specified period and can be read by anyone. In this process, opinions are submitted by the Minister of the Environment, the competent ministers, prefectural governors and mayors, and the public.

3.4.2 Environmental Mitigation

The Environmental Impact Assessment Law, in Article 1, stipulates that "measures should be taken to reflect the results of the environmental impact assessment in decisions related to the content of the project, such as measures for the conservation of the environment related to the project, to ensure that appropriate consideration is given to the conservation of the environment related to the project." This chapter calls for conservation measures against the impact on the environment caused by the project. In considering environmental conservation measures, priority is given to the avoidance or reduction of environmental impact. In the event of environmental damage, measures are required to compensate from the perspective of environmental conservation, such as the creation of a similar environment. Such environmental conservation measures have been pioneered and institutionalized in the West. They are called environmental mitigation [20]: (1) Avoidance means mitigating an aquatic resource impact by selecting the least-damaging project type, spatial location, and extent compatible with achieving the purpose of the project. Avoidance is achieved through an analysis of appropriate and practicable alternatives and a consideration of impact footprint. (2) Minimization means mitigating an aquatic resource impact by managing the severity of a project's impact on resources at the selected site. Minimization is achieved through the incorporation of appropriate and practicable design and risk avoidance measures. (3) Compensatory mitigation means mitigating an aquatic resource impact by replacing or providing substitute aquatic resources for impacts that remain after avoidance and minimization measures have been applied and is achieved through appropriate and practicable restoration, establishment, enhancement, and/or preservation of aquatic resource functions and services. They are sometimes referred to as avoidance, minimization, and compensatory mitigation.

In Japan's environmental conservation measures, when the habitat of endangered species is reduced due to a certain development project, it is common practice to secure an environment that can become a habitat in adjacent places, move the relevant species there, monitor the progress, and compile a post-report. However, in foreign environmental mitigation procedures, no net loss is required. This means that the quality (for example, the environment required for the habitat of a particular endangered species) and quantity (usually area) of a specific environment do not decrease before and after the project. If the quality of the environment to be secured is lower than the quality of the environment to be lost, it may be required to secure an area two or several times the area of the lost environment. In Japan's environmental assessment procedures, no net loss is not required. In the United States, the mechanism of environmental mitigation was born to conserve wetlands and prevent further reduction. To achieve no net loss in projects, a method (Habitat Evaluation Procedure; HEP) has been developed to quantitatively evaluate wildlife habitats [21]. With this quantification, it is possible to evaluate the loss and restoration of the environment in development projects and the results of natural restoration and conservation activities. The results are acknowledged and a mitigation banking system enables businesses that are required to compensate for development to purchase them [22]. Citizen groups promoting nature conservation can also raise funds by selling credits. Carbon offsetting is widely known in environmental finance, but mitigation banking has a more extended history. In recent years, following the model of carbon offsetting, the term biodiversity offsetting has come into use. Since the tenth Conference of the Parties (COP10) to the CBD was held in Nagoya in 2010, there has been ongoing debate about the necessity of establishing an international mechanism for biodiversity offsetting. However, this has not yet been achieved.

3.4.3 Eco-Roads

Transportation infrastructure, such as roads and railways, due to their linear form, has the characteristic of fragmenting the habitats of living organisms. This not only reduces the size of habitats, but also causes various impacts, including roadkill when animals attempt to cross and collide with cars or vehicles. It also fragments migration routes for organisms like amphibians that move between forests and watersides in their lifecycle. Furthermore, it results in a decrease in genetic diversity due to the inability to interbreed with neighboring groups of the same species (known as populations). For instance, according to the Ministry of the Environment's Tsushima Wildlife Conservation Center, between 1992 and 2021, 117 Tsushima leopard cats, which are endemic to Tsushima and are also designated as natural monuments of the country, have perished in traffic accidents. The current population is estimated to be approximately 90–100 individuals, indicating that reducing traffic accidents is crucial for the survival of the species.

In order to mitigate the aforementioned impacts on organisms to the greatest extent possible, roads and railways are being constructed with consideration for animal movement, collectively referred to as eco-roads. Various cases have been accumulated in the West and introduced to Japan, in a manner analogous to environmental impact assessments. The creation of eco-roads was first advocated in the 11th Five-Year Road Improvement Plan in 1993 [23]. The specific measures are indeed diverse and not only animal species are targeted; eco-roads that consider plants are also being built. In the West, eco-roads are often positioned as part of the environmental mitigation process mentioned earlier [23]. In Japan too, various examples have been seen, but road construction is often decided first, tunnels or bridges for animal movement are installed as a remedial measure, and many of them are questionable in terms of their effectiveness as eco-roads. Additionally, it is estimated that between 89 million and 340 million wild birds die annually in collisions with cars across the United States [24], indicating the necessity for further measures in the future.

3.4.4 Invasion of Alien Species

The clover, now commonly observed in Japanese grasslands, was originally an alien species. It is a well-documented phenomenon that spread in Japan after being packed in the United States as cushioning material for goods when trade between America and Japan became active. Along with human migration, various organisms are intentionally or unintentionally transported. The invasion of alien species by transportation is a significant issue for biodiversity conservation. The Ministry of the Environment has identified four crises threatening Japan's biodiversity in the National Biodiversity Strategy [25]. One of these is the crisis caused by things brought in by humans, which includes alien species.

Japan does not border any other country, so the invasion of alien species is mostly by ships. For example, one that has been attracting attention in recent years is the red imported fire ant. The red imported fire ant (*Solenopsis invicta*) is an ant native to central South America that was first confirmed in Japan in 2017. It is believed to have been transported by ship hidden in international cargo. Red imported fire ants affect ecosystems and cause human harm because they are poisonous. Therefore, they have been designated as an alien species requiring urgent measures, which is necessary to prevent their invasion and establishment.

A typical example related to ships is the case of organisms being transported mixed in ballast water. Ballast water is needed, especially when the cargo load is small on cargo ships, to stabilize the ship's operation. It is common practice to load seawater into the ballast tank. Many aquatic organisms mix into this seawater. Seawater is discharged when loading cargo at different ports so that it can be transported over considerable distances. It has been observed that the survival rate of mixed aquatic organisms is increasing with the speeding up of ships. A well-known example in Japan is the Mediterranean mussel, which is designated as an alien species that requires caution. Of course, native Japanese organisms are also transported, and sea mustard is increasing in Australia, causing significant damage to the fishing industry.

The increase in air travel demand is also promoting the movement of organisms and bringing risks to humans. The most prevalent unwelcome guests transported by aircraft are pathogens and viruses. In 2003, it was discovered that travelers using airplanes were infected with severe acute respiratory syndrome (SARS), prompting countries, including China, to implement emergency responses. Additionally, in 2003, West Nile fever spread in North America, and it was highlighted that the spread of infection could be facilitated by mosquitoes on airplanes. In light of these developments, the Ministry of Health, Labor and Welfare issued a notice on "Guidance on measures against West Nile fever vector mosquitoes for aircraft coming from North America." The movement of people and things has become globalized, and the risk of a pandemic of infectious diseases has been highlighted on numerous occasions. The first pandemic to emerge was the 2019 coronavirus disease (COVID-19).

3.4.5 Nature Positive

The Kunming-Montreal Global Biodiversity Framework, which was agreed upon at COP15 in December 2022, outlines an urgent mission to take immediate action to halt and reverse biodiversity loss. This action is necessary to put nature on a path to recovery for the benefit of people and the planet. In order to achieve this goal, it is essential to conserve and sustainably use biodiversity, as well as to ensure the fair and equitable sharing of benefits from the use of genetic resources. Furthermore, it is necessary to provide the necessary means of implementation up to 2030. This aligns with the concept of Nature Positive, as proposed by Locke et al. [26] This means achieving zero net loss of nature from 2020, net positive by 2030, and full recovery by 2050 (Fig. 3.3). To achieve the 2050 goal, it is necessary to continue maintaining and restoring ecosystems until there are enough ecosystems to support all life on Earth, including future generations, and to protect the stability and resilience of the Earth system. Consequently, it is feasible to achieve the 2050 vision of the CBD, "Living in Harmony with Nature," the 2050 carbon neutral goal of the

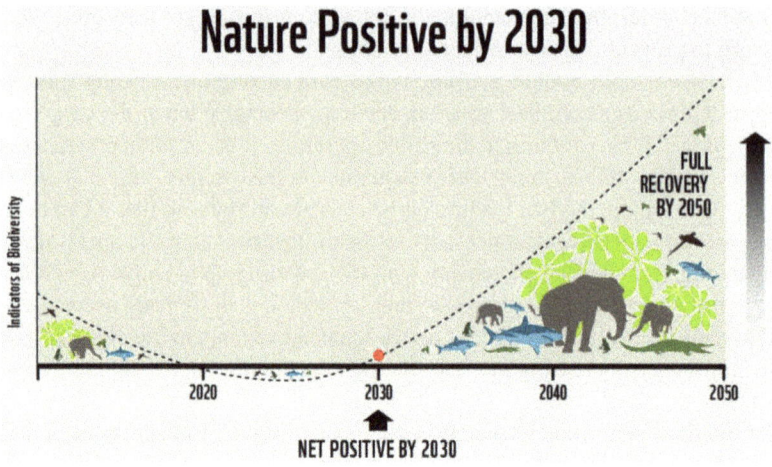

Fig. 3.3 The trajectory of nature positive by 2030 [26]

UNFCCC, and the SDGs [26]. Prior discussions have centered on methods to minimize the impact on ecosystems and biodiversity, or to achieve zero impact through mechanisms such as environmental mitigation. This represents a novel approach, as it is the first time a goal to bring nature into the positive has been agreed upon. The 2030 target of the Kunming-Montreal Biodiversity Framework, in addition to the initially introduced 30 by 30, includes a goal to place 30% of degraded ecosystems under effective recovery. The latest national biodiversity strategy of Japan, revised in response to COP15, also includes Nature Positive in the subtitle, echoing international trends.

3.5 Transformation of Corporate Management Promoted by Environmental Issues

As environmental issues become increasingly pressing, private companies are increasingly bound by environmental regulations set by governments and local public bodies. This has often led to the development of new technological innovations. It has also become common for companies to engage in activities that lead to environmental improvements as part of their Corporate Social Responsibility (CSR). However, the pressing global environmental issues are demanding even greater changes from private companies, namely the transition to natural capital management. Natural capital is defined as the world's stock of natural resources, which includes geology, soils, air, water, and all living organisms. The flow produced from natural capital can be seen as ecosystem services. Properly valuing and managing the value of natural capital can enhance the sustainability of corporate management.

Investments are starting to be made in companies that incorporate this concept of natural capital into their management.

In 2015, the Task Force on Climate-Related Financial Disclosures (TCFD) was established, requiring companies to provide information to investors about their efforts to mitigate the risks of climate change and to be transparent about their governance practices. This means that companies must address climate change in order to receive investment. In the TCFD report [27], the transportation group encompasses the industries of air cargo transportation, air passenger transportation, shipping, rail transportation, truck services, and automobiles/parts. It is likely to face financial challenges from two main factors. One is that policymakers are setting stricter targets for the emissions and combustion efficiency of transport operators. The other is that new technologies for low-emission, fuel-efficient transport equipment are causing changes in the competitive and investment environment. In light of these considerations, the manner in which information pertaining to climate change is disclosed is elucidated.

Similarly to climate change, the Taskforce on Nature-related Financial Disclosure (TNFD) was established in 2021 for the natural environment, and its framework was published in September 2023. The TNFD inherits the structure of the TCFD, and of the 14 items recommended for disclosure, 11 are required to be disclosed by the TCFD. One of the items added by the TNFD is the priority area. In contrast to GHGs, which affect global climate change in a uniform manner regardless of their point of origin, the significance of a given impact varies by region in the natural environment. In order to fulfill the disclosure requirements, information must be made available in areas that meet both of the following conditions: (1) regions where significant nature-related impacts and risks are considered to exist, and (2) regions where ecosystem degradation is significantly progressing [28].

The advent of the TCFD and TNFD has made it imperative for all companies to integrate climate change measures and natural environment conservation into their management. Failure to address global environmental issues may jeopardize the survival of the company. Conversely, the rapid spread of BEVs and mobility-sharing services overseas in recent years demonstrates that global environmental issues are creating new business opportunities.

3.6 Conclusion

The history of transportation and the environment has been marked by various environmental issues arising with the development of modern transportation and technology. In response, governments and others have been strengthening regulations related to the environment. However, global environmental issues such as climate change and loss of biodiversity are casting a large shadow over the survival of our humanity. In light of this, societal changes unlike any before are being demanded in order to achieve goals such as carbon neutrality and nature positivity. Transportation is no exception. The global restrictions imposed by the COVID-19 pandemic provided a unique opportunity to assess the impact of mobility restrictions on human well-being. While the ability to hold virtual meetings and take online classes was a positive outcome, the experience also highlighted the significant stress that can be caused by limiting the freedom of movement.

In 2020, the global economy and social activities declined due to the pandemic, resulting in a 6.4% reduction in carbon dioxide emissions (2.3 billion tons) [29]. In order to achieve the temperature rise target of below 1.5 °C, as outlined in the Paris Agreement, it is necessary to reduce emissions by 7.6% annually until 2030. A 6.4% reduction is insufficient. It is evident that the societal changes being demanded of us are significant. It may be argued that we are entering an era where the environment will determine how transportation is, rather than transportation affecting the environment.

References

1. Meadows DH (1972) The limits to growth: a report for the Club of Rome's project on the predicament of mankind. Universe Books, New York, p 205
2. Rockström J, Steffen W, Noone K, Persson Å, Chapin FS, Lambin EF, Lenton TM, Scheffer M, Folke C, Schellnhuber HJ, Nykvist B, de Wit CA, Hughes T, van der Leeuw S, Rodhe H, Sörlin S, Snyder PK, Costanza R, Svedin U, Falkenmark M, Karlberg L, Corell RW, Fabry VJ, Hansen J, Walker B, Liverman D, Richardson K, Crutzen P, Foley JA (2009) A safe operating space for humanity. Nature 461(7263):472–475
3. Steffen W, Richardson K, Rockstrom J, Cornell SE, Fetzer I, Bennett EM, Biggs R, Carpenter SR, de Vries W, de Wit CA, Folke C, Gerten D, Heinke J, Mace GM, Persson LM, Ramanathan V, Reyers B, Sorlin S (2015) Sustainability. Planetary boundaries: guiding human development on a changing planet. *Science* 347(6223):1259855
4. Richardson K, Steffen W, Lucht W, Bendtsen J, Cornell SE, Donges JF, Drüke M, Fetzer I, Bala G, von Bloh W, Feulner G, Fiedler S, Gerten D, Gleeson T, Hofmann M, Huiskamp W, Kummu M, Mohan C, Nogués-Bravo D, Petri S, Porkka M, Rahmstorf S, Schaphoff S, Thonicke K, Tobian A, Virkki V, Wang-Erlandsson L, Weber L, Rockström J (2023) Earth beyond six of nine planetary boundaries, *science*. Advances 9(37):eadh2458
5. IPCC: AR6 climate change 2021—the physical science basis, 2021
6. WMO: Global temperatures set to reach new records in next five years, 2023. https://public.wmo.int/en/media/press-release/global-temperatures-set-reach-new-records-next-five-years. Accessed on 29 Oct 2023
7. IPBES: summary for policymakers of the global assessment report on biodiversity and ecosystem services (summary for policy makers), Bonn, 2019
8. Ruckert A, Zinszer K, Zarowsky C, Labonte R, Carabin H (2020) What role for one health in the COVID-19 pandemic? Can J Public Health 111(5):641–644
9. OECD: transport and the environment, Paris, pp. 131, 1988
10. OECD: Transport and the environment, Japan Economic Research Council, pp. 212, 1993. (Japanese version)
11. National Institute for Environmental Studies: Explanation of Environmental Technology—Automobile Recycling Technology, 2023. https://tenbou.nies.go.jp/science/description/detail.php?id=67. Accessed on 29 Oct 2023. (in Japanese)
12. European Environment Agency: landscape fragmentation in Europe—Joint EEA-FOEN report, Luxembourg, Publications office of the European Union, pp. 87, 2011
13. IEA: CO2 emissions in 2022, IEA Publications, pp. 17, 2023

14. Takeshi Momoda: (Latest 2023) What is the prevalence of EVs? Explanation of EV situation in Japan and the world, 2023. https://evdays.tepco.co.jp/entry/2021/09/28/000020. Accessed on 29 Oct 2023. (in Japanese)
15. Holland SP, Mansur ET, Muller NZ, Yates AJ (2016) Are there environmental benefits from driving electric vehicles? The importance of local factors. Am Econ Rev 106(12):3700–3729
16. Moreno C, Allam Z, Chabaud D, Gall C, Pratlong F (2021) Introducing the "15-Minute City": sustainability, resilience and place identity in future post-pandemic cities. Smart Cities 4(1):93–111
17. Li S, Jianwei X, Yang L, Zhang F (2020) Transportation and the environment—a review of empirical literature. World Bank Group, p 55
18. Li S, Liu Y, Purevjav A-O, Yang L (2019) Does subway expansion improve air quality? J Environ Econ Manag 96:213–235
19. Ministry of the Environment: Environmental impact assessment in Japan, Environmental Impact Assessment Division Minister's Secretariat, Ministry of the Environment, pp. 17, 2023
20. United States Environmental Protection Agency: Types of mitigation under CWA Section 404: avoidance, minimization, and compensatory mitigation. https://www.epa.gov/cwa-404/types-mitigation-under-cwa-section-404-avoidance-minimization-and-compensatory-mitigation. Accessed on 13 May 2024
21. Schamberger M, Krohn WB (1982) Status of the habitat evaluation procedures. US Fish & Wildlife Publications 48
22. Weems WA, Canter LW (1995) Planning and operational guidelines for mitigation banking for wetland impacts. Environ Impact Assess Rev 15(3):197–218
23. Overseas Eco-Road Research Team: Eco-Road Book—Collection of Overseas Examples of Road Construction Coexisting with Living Creatures, pp. 124, 1999. (in Japanese)
24. Loss SR, Will T, Loss SS, Marra PP (2014) Bird–building collisions in the United States: estimates of annual mortality and species vulnerability. Condor 116(1):8–23
25. Ministry of the Environment: National Biodiversity Strategy 2023–2030—Roadmap towards Realizing a Nature-Positive Society, pp. 217, 2023
26. Locke, H., Rockström, J., Bakker, P., Bapna, M., Gough, M., Hilty, J., Lambertini, M., Morris, J., Polman, P. and Rodriguez, C. M.: A nature-positive world: the global goal for nature, 2021
27. TCFD: Implementing the recommendation of the task force on climate-related financial disclosures, pp. 82, 2017
28. TNFD: Recommendations of the taskforce on nature-related financial disclosures, pp. 153, 2023
29. Meinhardt J, Radke J, Dittmayer C, Franz J, Thomas C, Mothes R, Laue M, Schneider J, Brunink S, Greuel S, Lehmann M, Hassan O, Aschman T, Schumann E, Chua RL, Conrad C, Eils R, Stenzel W, Windgassen M, Rossler L, Goebel HH, Gelderblom HR, Martin H, Nitsche A, Schulz-Schaeffer WJ, Hakroush S, Winkler MS, Tampe B, Scheibe F, Kortvelyessy P, Reinhold D, Siegmund B, Kuhl AA, Elezkurtaj S, Horst D, Oesterhelweg L, Tsokos M, Ingold-Heppner B, Stadelmann C, Drosten C, Corman VM, Radbruch H, Heppner FL (2021) Olfactory transmucosal SARS-CoV-2 invasion as a port of central nervous system entry in individuals with COVID-19. Nat Neurosci 24(2):168–175

Open Access This chapter is licensed under the terms of the Creative Commons Attribution-NonCommercial-NoDerivatives 4.0 International License (http://creativecommons.org/licenses/by-nc-nd/4.0/), which permits any noncommercial use, sharing, distribution and reproduction in any medium or format, as long as you give appropriate credit to the original author(s) and the source, provide a link to the Creative Commons license and indicate if you modified the licensed material. You do not have permission under this license to share adapted material derived from this chapter or parts of it.

The images or other third party material in this chapter are included in the chapter's Creative Commons license, unless indicated otherwise in a credit line to the material. If material is not included in the chapter's Creative Commons license and your intended use is not permitted by statutory regulation or exceeds the permitted use, you will need to obtain permission directly from the copyright holder.

Chapter 4
Public Transportation and the Local Community

Fumihiko Nakamura, Ryo Ariyoshi, and Yurie Toyama

This chapter discusses public transportation. Considering the somewhat complex definition of public transportation, this chapter has been set up in accordance with the purpose of this book, starting with the definition of public transportation and delving into the issues in the local community in the future.

4.1 Introduction

Local life is based on the movement of people and goods, most of which use roads. Originally, it was walking, then moving with animals such as horses, and then the development of transportation tools such as bicycles and cars.

In such a context, there are services for moving when one cannot operate a vehicle, and services that bundle and transport people's movements when there is a concentration of mobility needs. These are what we call public transportation. In recent years, services that lend transportation tools like shared bicycles and services that share the time and space of transportation tools like ride-sharing have also emerged.

Public transportation plays a role in supporting and improving the efficiency of movement in the local community, and it is also indirectly and directly related to traffic safety. The operation by professional drivers increases safety, and especially for the elderly and others with reduced physical abilities, there is also the effect of reducing the risk of traffic accidents because they do not have to drive themselves.

F. Nakamura (✉)
Graduate School of Frontier Sciences, The University of Tokyo, Bunkyō, Japan
e-mail: nakamura-fumi@edu.k.u-tokyo.ac.jp

R. Ariyoshi · Y. Toyama
Institute for Future Society Creation, Nagoya University, Nagoya, Japan

In this chapter, we focus on the relationship with the local community, and do not cover services that handle long-distance travel of several hundred kilometers or more, such as high-speed railways, express buses, ferries, and airplanes. The main target is road services that handle daily travel. One of the main reasons is that these services are more deeply involved in road traffic safety.

4.2 Basics of Public Transportation

4.2.1 Definition

What can be called public transportation on the road can be organized as shown in Fig. 4.1 as the definition of public transportation in Europe seen in the International Association of Public Transport (UITP) and others. In Japan and some other countries, public transportation and transportation business are synonymous in terms of policy. In this book, we distinguish the two as internationally applicable concepts. The figure also distinguishes between transportation business and public transportation as separate concepts.

We classify those easy-to-access mobility services as public transportation, which anyone can access casually on a daily basis, even if there may be the hassle of paying fares or registering as a member, without having to go through any particularly complicated reservation procedures.

On the one hand, buses chartered for group activities such as school excursions are classified as transportation businesses in Japan, though they are buses from the mechanical vehicle's aspect and though the bus operators provide the service. However, they will not be included in public transportation in this context. On the other hand, bike-sharing services, for example in Japan, are considered rental businesses, but from the user's perspective, they are convenient mobility services and will be classified as public transportation. Many services seen in the Global South, such as shared rides in small vehicles or taxi-like services, are referred to as

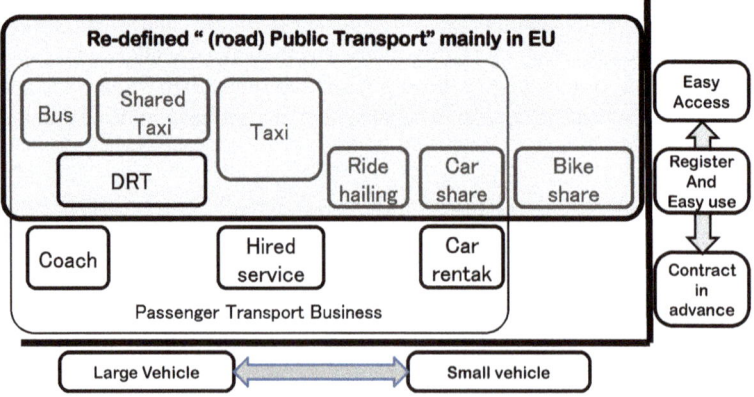

Fig. 4.1 Definition of public transportation

4 Public Transportation and the Local Community

paratransit or intermediate public transportation. Based on the above definition, they will be classified as full-fledged public transportation in this book.

4.2.2 Current Situation and Challenges

The challenges faced by public transportation in communities vary by country and region, but they can be organized into common points such as the division of roles between the public and private sectors, planning, operation, division of operation, division of competitive and cooperative areas, how to bear the operating costs, and ensuring safety and public order. The demand for public transportation is for everyday activities, which are derived demands associated with some activities in the region. Therefore, they should be planned in conjunction with the way regional activities, land use, and spatial composition are. In this part, it is required that the division of roles between the public and private sectors in the planning and operation stages is clearly organized. Especially in urban planning, where the administration guides and regulates with a focus on residents, there are situations where everything cannot be left to the private sector. Considering that regional activities are guaranteed by the mobility provided by mobility services, it is not unnatural for the region to bear the cost of mobility services. A model that covers all costs with only the fare income from users would be rather unnatural. However, in reality, there are many mobility services that have naturally evolved from unique businesses in a competitive environment by the private sector, and it is often not easy to introduce a management system by the government or administration later. The utilization of private sector vitality through deregulation has been advocated and practiced in various situations. Although it is a field with various discussions, there are increasing situations where it cannot be said that it is enough to just deregulate.

4.2.3 Noteworthy Main Urban Transportation Methods

Over the past 20 years or more, words such as LRT (light rail transit: light rail, next-generation tram) and BRT (bus rapid transit: bus rapid transit system) have been used in policy documents in various countries. Since their definitions are somewhat unclear, they will be organized here.

LRT
LRT is a system that largely follows the vehicle standards of trams, runs its route and tracks, in many cases, through the central part of wide roads. While the tracks are generally double-tracked, they can become single-tracked depending on the road conditions. Depending on the situation, it flexibly responds with dedicated tracks, elevated running, underground running, direct connections with conventional railway lines, etc.

The perspectives that differentiate traditional trams and LRTs include, in addition to the flexibility of the above-mentioned routes, the use of modern vehicles with low floors, large transport capacity, high acceleration and deceleration performance, and high maximum speed, the ingenuity in service content such as fare payment methods, consideration for coordination with other means of transportation, and consideration for coordination with urban development such as transit malls.

BRT

Basically, it is a route bus using bus vehicles, but it refers to those that are significantly differentiated from conventional route buses as a whole by realizing mass transportation, high punctuality, and high-speed (rapid) transportation through various ingenuities. It is not appropriate to call it BRT just by introducing articulated vehicles. The "high speed" mentioned here means faster than vehicles in traffic on congested roads. There is often an image of a BRT bus running fast in a scene where there is a bus-only lane separated by curbs in the central part of a wide road, and other lanes are congested with general vehicles.

There are currently three major trends in BRT. One is the railway alternative type of BRT, such as in the Federal City of Curitiba, Brazil, and Bogota, Republic of Colombia. These have given up on the construction of subways, etc., due to cost and construction period and are aiming for transportation capacity and speed comparable to subways.

BHNS (BHLS in English) can be seen in many cases in France. Here, they were considering the introduction of LRT, but gave up on LRT due to reasons such as introduction space, cost, and construction period. In that sense, comparing this type of BRT and LRT is not very meaningful. Apart from the details, if there is money and physical space allows, LRT is good, and if that is not possible, BRT is good.

Another one, unique to Japan, is the railway discontinued line BRT. This is not high-speed or mass transportation, but after a railway line that has been familiar to locals for many years was discontinued due to a natural disaster and no prospect of recovery, the railway track part is paved, and after introducing bus-only road regulations, only buses are allowed to run. Although it is not necessarily high-speed or mass, the significance as a railway alternative is heavy.

4.3 Public Transportation and Urban Activities

4.3.1 Introduction

In this section, we will discuss the relationship between the activities and travels of urban residents, the challenges related to mobility in cities and approaches to solving them, the reconstruction of mobility systems as a means of achieving regional goals, co-creation activities to realize the desired regional vision, and the importance of human resource development as a driving force for these activities.

4 Public Transportation and the Local Community

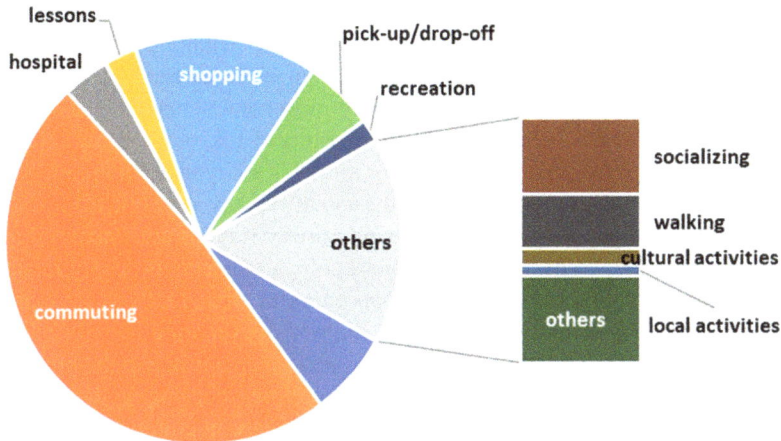

Fig. 4.2 Composition of weekday travel purposes

4.3.2 Travels and Activities of People

As already mentioned, most travels of people are derived from their needs. People engage in various activities such as work, learning, medical visits, shopping, eating, and leisure at appropriate or designated places to meet their diverse needs, and travel is necessary to reach each of these places. Figure 4.2 shows the composition of weekday travel purposes for residents of the Tokyo metropolitan area. The majority of travels are derived for commuting, shopping, and other purposes, while the proportion of intrinsic travels, such as walking or exercising, where the travel itself is the purpose of the activity, is small. Travel as a derived need is always integrated with the activity at the destination, and to understand people's travels in cities, it is necessary to understand people's activity needs and the situation of opportunities that can satisfy these needs. Opportunities refer to places where individuals can obtain the utility they expect, such as offices for work and stores for purchasing goods and services, and in a broader sense, it includes virtual places like online meetings. Therefore, the situation of opportunities refers to the combination of the state of "density" of how many available options an individual has, the state of "quality" of how well each opportunity can satisfy the individual's needs, and the state of "cost" of how much the individual has to pay to obtain the opportunity.

4.3.3 Reconstruction of Regional Transportation

As mentioned in the previous subsection, many of the opportunities to satisfy the needs of urban residents are concentrated in the city center. For example, in Japan, according to the national census (2015), about 40% of the daytime population (approx. 7.63 million people) of the central Tokyo, or about 3.18 million people, are inflows from outside, with the population of the three neighboring prefectures of Kanagawa, Saitama, and Chiba accounting for about 80%, indicating a large amount of travel as a derived need. The source of such movement demand is the residence of the traveler, and in the metropolitan area consisting of Tokyo, Kanagawa, Saitama, and Chiba, suburban residential areas along railroads are typical. In the metropolitan area, urban areas are formed along the radial railway network from the city center, and residential areas spreading around railway stations are connected to the stations by bus routes. The daily life of metropolitan residents is deeply related to the public transportation network and the opportunities around railway stations, and the modal share of public transportation is quite high even on a global scale. The high share of public transportation in the metropolitan area is due to the high use of railways for commuting to work and school. Therefore, focusing on those aged 65 and over, who are expected to see a significant decrease in commuting needs, the share of public transportation remains flat, and the share of cars is larger and increasing (Fig. 4.3).

According to the Tokyo Metropolitan Area Person-Trip Survey (2018), the number of trips per person per day has decreased for the first time since the survey began in 1968. The decrease in travel for business or private activities, especially the decrease in outings by those are not commuting to work or school, might be considered to have a significant impact. It is speculated that the availability of private cars is related to the decrease in the number of trips of these groups. Compared to those who own a private car that they can drive and use freely, those who cannot use it for reasons such as not having a driving license or a vehicle have smaller number of trips per day, especially among those aged 65 and over, the difference is noticeable (Fig. 4.4).

Ensuring sustainable mobility to meet activity needs other than commuting in suburban residential areas of large metropolitan areas where rapidly aging is an urgent issue. Providing alternative travel modes such as walking, cycling, and

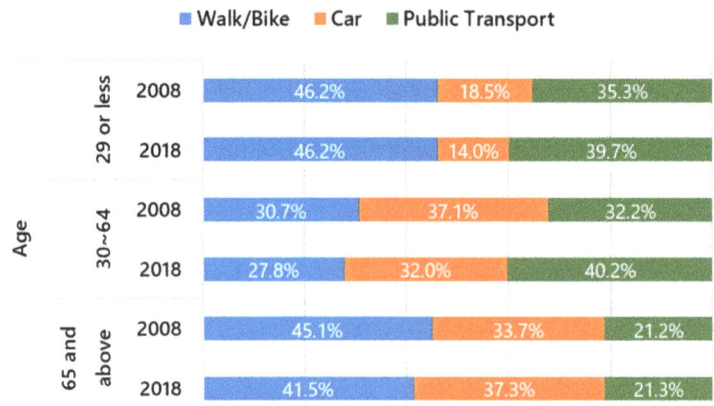

Fig. 4.3 Modal split by age class

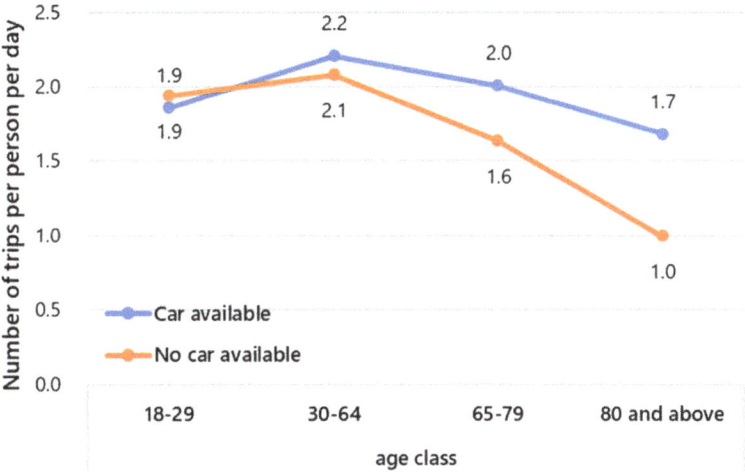

Fig. 4.4 Average number of trips by age class

private cars for the elderly, who are prone to mobility decline due to aging, can suppress the decrease in outing frequency, maintain opportunities for interaction and social participation, extend individual healthy life expectancy through these, maintain social ties, and ultimately lead to a reduction in social security-related expenditures. Under such a concept, it is essential to redesign the vision of the region and reconstruct the regional mobility system that can contribute to its realization. An approach that utilizes all mobility-related resources (vehicles, services, personnel, know-how, data, etc.) existing in the region based on conventional mobility services such as railways, buses, and taxis, redefines other mobility services included in the scope of public transport as necessary, and designs, evaluates, and demonstrates policy alternatives while repeating dialogues, and runs the PDCA cycle agilely is considered effective.

4.3.4 Co-creation and Human Resource Development in a Region

In this section, we introduce a case in Japan where academia, industry, and government are collaborating to implement new mobility services, with the aim of improving the value and sustainability of residential areas. A private railway operator (also developer) is exploring ways to secure mobility that meets the needs of residents and the vision of the region through demonstration experiments of small-scale shared mobility services. Gradually involving various regional entities, measures are being implemented that integrate with town management, such as the establishment of local bases and community formation. This is a case that suggests the importance of co-creation in connecting mobility service planning to implementation, and the importance of human resource development as its driving force.

The Tomioka district in Kanazawa ward, Yokohama city, is a hilly residential area with a population of about 16,000, spreading to the west of Keikyu Tomioka station on the Keihin Express Railway. This area, where land development and subdivision have been promoted by Keikyu Corporation since the late 1960s, is home to many people who commute to the city center of Yokohama and Tokyo by train, and houses, mainly single-family homes, line up along the continuous slopes. In 2018, Keikyu had signed an agreement with Yokohama city on urban development and was working to solve the problems of residential areas along the line and then signed a cooperation agreement with Yokohama National University in the field of mobility; discussions on mobility services that contribute to the sustainability of the Tomioka district have begun. In this project, a rideshare service using a small low-speed electric vehicle was positioned as a role to support short-distance travel between residents' homes and bus stops and local facilities, with the aim of improving the overall mobility of the area by complementing existing buses and taxis. This small-scale shared mobility service was named "Tomio-Cart" by students at Yokohama National University. The specifications and results of the demonstration experiment of Tomio-Cart, which has been ongoing for five years from 2018 to the present, are organized in Table 4.1. This initiative, which started with the practical use of small low-speed electric vehicles, gradually expanded the verification area while gaining the understanding and cooperation of various entities, and it reached a paid experiment that combined fixed route and demand responsible operation in 2020. In the demonstration experiment for 2022, the demand was increasing steadily, and by September 2023, the number of rides per day per vehicle has grown to about 2.4 times that of the start month. This level is about twice the value during the free experiment period in 2020, and the long-term operation of Tomio-Cart over a year suggests that it is gradually increasing its users and taking root in the community.

The Tomio-Cart Project has been continuously running a PDCA cycle for five years, testing various forms of mobility services in one field, making decisions based on demonstration results, taking actions for policy improvement, and planning for the next phase. The biggest factor making this possible is believed to be that Keikyu—the railway, bus, and taxi operator and also developer of the Tomioka area—is leading this project. They see the purpose of this project not as the introduction of a new transportation system, but as improving the value of residential areas along the line. They hope that by establishing and sustaining mobility support services like Tomio-Cart, which connects hillside residential areas and stations, the outings of residents, mainly the elderly, will be maintained and increased, and the management of railway operations, station-front commerce, and local real estate businesses will continue. In this sense, for Keikyu, maintaining Tomio-Cart is not a supplement to the feeder travel modes of a specific railway station, but an investment in all of its own line areas. Also, being able to build a collaborative team with local governments and local universities from the initial stage as counterparts to share the vision of such suburban areas of large cities was considered a major factor

4 Public Transportation and the Local Community

Table 4.1 Summary of demonstration experiments

	1st stage		2nd stage							2022		
			2020 (Oct to next Feb)						2021 (Nov to next Jan)	Starting month (Dec '22)	6 months later (Jun '23)	9 months later (Sep '23)
			Fixed route service				DRT^a service					
	2018 (Oct–Nov)	2019 (Nov–Dec)	Free period	Paid period	Free period	Paid period	Free period	Paid period				
Operating days	20	18	52	23	62	23	75			16	17	16
Day of operation	Weekdays / Sat/Sun	Weekdays / Sat/Sun	Weekdays / Sun	Weekdays / Sun	Weekdays / Sun	Weekdays / Sun	Weekdays / Sat			Mon/Tue/Wed/Fri (excluding holidays)		
Operating hours	9:00~17:00	9:30~17:30	9:00~17:00	9:00~17:00	9:00~17:00	9:00~17:00	8:30~18:00			10:00~19:30		
Service interval	Approx. 30 min	Approx. 20 min	Approx. 30 min	Approx. 30 min	–	–	Approx. 50 min			Approx. 50 min		
reservation	Unnecessary	Unnecessary	Unnecessary	Unnecessary	Required (15 min ago)	Required (15 min ago)	Unnecessary			Unnecessary		
Operating route	Fixed	Fixed	Fixed	Fixed	Variable	Variable	Fixed			Fixed		
Boarding/ Alighting point	Fixed	Fixed	Any	Any	Fixed (many)	Fixed (many)	Any			Any		
Number of vehicles in operation	Capacity 3×1	Capacity 3×1	Capacity 3×2	Capacity 3×2	Capacity 4×2	Capacity 4×2	Capacity 8×1			Capacity 5×1		
		Capacity 6×1			Capacity 3×1	Capacity 3×1						
Vehicle type	Electric cart	Electric cart	Electric cart gasoline car	Electric cart gasoline car	Gasoline car	Gasoline car	Gasoline car			Gasoline car		

(continued)

Table 4.1 (continued)

	1st stage	2nd stage						
			2020 (Oct to next Feb)					2022
	2018 (Oct–Nov)	2019 (Nov–Dec)	Fixed route service		DRT* service		2021 (Nov to next Jan)	Starting month (Dec '22)	6 months later (Jun '23)	9 months later (Sep '23)
			Free period	Paid period	Free period	Paid period				
Fare (adult)	Free of charge	Free of charge	Free of charge	200 yen/ride	Free of charge	300 yen/ride	200 yen/ride	200 yen/ride		
Number of registrants	123	258	1870				–	–		
Number of unique users	67	78	336	Not clear	342	43	Not clear	Not clear		
Total number of rides	124	247	1295	336	1607	105	1552	265	527	626
Number of rides per vehicle per day	6.2	6.8	13.8	7.3	8.6	1.5	20.7	16.6	31.0	39.1
Operating cost per ride	0.16D yen	0.14D yen	0.08D yen	0.14D yen	0.12D yen	0.66D yen	0.05D yen	0.06D yen	0.03D yen	0.03D yen

in being able to continue the project over the long term. In May 2021, Keikyu and Yokohama city took on the role of facilitators, and the town development guideline was published. This guideline is established through dialogue with residents in many workshops. Based on this guideline, Tomio-Cart will increase the local activity needs of residents while collaborating with the creation activities of local bases, aiming to utilize it as a travel mode for such activities. The " Okamachi Living" opened in front of the Keikyu Tomioka station in June 2023 is a multi-purpose hub space for the community, thanks to the generosity of the building owner. It is also used as a waiting room for Tomio-Cart users, and they can wait for Tomio-Cart in a comfortable space while watching the expected passing time and current location information of the vehicle on the digital signage installed. In addition to short stays for breaks before returning home and organizing luggage, it can be used for various activities such as self-study, telework, and workshops by residents and is being utilized as a new community base. By creating such a place, the frequency of outing by residents increases, and Tomio-Cart aims to be used more by diverse generations as a travel mode for this purpose. Considering the above, one of the issues in mobility planning policy in hilly residential areas centered around railway stations is the way in which various entities collaborate to achieve a sustainable mobility system that contributes to the realization of the regional vision, and the methodology for demonstrating the value of supporting travels as derived demands, from multiple perspectives. Research on theories and methods for objectively and reliably evaluating how the creation of places like the aforementioned "Okamachi Living" brings about changes in the frequency of residents' outings, choice of destinations and travel mode, and the impact on individual health and subjective well-being, and how it affects the strengthening of social cohesion and economic revitalization in the region, is necessary. Without such academic backing, it would be difficult to continue collecting operating funds for regional mobility services like Tomio-Cart from beneficiaries other than passengers. Also, the problem of suburban areas of large cities centered on dependence on private cars cannot be solved by single technology development such as autonomous driving and on-demand transportation, and it is important to recognize that comprehensive responses including land use planning are necessary. In that respect, in suburban areas, which are the source of economic activity in large cities, residents with advanced and specialized knowledge and skills live at high density, and there are universities working on solving social issues and companies with customers in the region. In the future, it will be important to create a movement to change the situation in the region in a more desirable direction by making use of the strength of having such diverse resources accumulated and to develop producer talents. In the suburbs of large cities, there are already regional activities involving various entities including citizens, physical spaces (including vacant facilities such as vacant houses and shops) to accommodate such activities, organizations and systems to support and manage activities and places, and knowledge and know-how accumulated in the region through various initiatives including regional events and demonstration experiments. However, these activities are often independent, and assets including knowledge and human resources are not shared, so a new mechanism to effectively link them is needed. One of the key entities to this is believed to be universities with ties to the region. As an institution for research and educational activities, universities can make a significant contribution by continuously interacting with the region from a neutral position, providing academic advice and technical support from a professional perspective as needed, and human support including students, and through such involvement in the region, a virtuous cycle of producing practical research results in the field of urban and mobility can be created. The strategic development of such regional co-creation mechanisms and the cultivation of the human resources that drive them will lead to the realization of sustainable regional mobility.

4.4 Public Transportation and New Technology

This section discusses public transportation and new technologies. Note that here public transportation mainly refers to urban land transportation (excluding airplanes, ships, and intercity railways).

Technology, which is constantly being innovated with each era, continues to have a significant impact on public transportation, evolving it to the public transportation we use today and will continue to develop in the future. Here, we will organize the technology related to public transportation in the modern era and how to deal with future technologies.

4.4.1 The Relationship Between Public Transportation and New Technology

Looking back at the relationship between new technology and public transportation historically, the development of the steam engine due to the industrial revolution in the late eighteenth century, and its application to steam locomotives, could be the start of modern public transportation. As urban transportation, the London Underground opened in 1863, running steam locomotives underground, and trams were put into practical use in Berlin in 1881, among other innovations and practical applications in urban transportation.

In the 1830s, steam buses were active in London, but the development of traffic regulations and the balance of operations were in an era of trial and error (e.g., the Red Flag Act/Locomotive Act).

Meanwhile, in America, Henry Ford launched the "Model T" in 1908, marking the beginning of the era of mass production of automobiles.

In the twentieth century, concerns regarding noise, exhaust gases, and traffic safety due to the progress of motorization, and the increase in population and urbanization in cities around the world, led to a demand for public transportation that is environmentally friendly, safe, and can transport more efficiently.

Today, awareness of environmental issues is given, and with the evolution of ICT technology, the evolution of public transportation continues in a direction that approaches the "experience" of movement by efficiently combining various information.

4.4.2 Expectations for New Technology in Public Transportation

The role required of public transportation changes with each era according to urban social issues and interests such as urban growth and decline, population increase and decrease, and interest in the environment. In our country, in general, in a society with a declining population, maintaining the service level necessary for life is required in rural areas. On the other hand, in major cities including Tokyo, even after the Covid-19, congestion relief during rush hours is required, and public transportation services that support the local economy and play a role as an international city are required. Behind these needs are the emergence of various social issues and changes in interest on a global scale, and public transportation is also required to respond to them.

To do this, it is essential to utilize new technologies. In modern public transportation, the motivation to utilize new technologies can be summarized into about five major points.

Firstly, there is "reducing environmental impact" and "achieving carbon neutrality." Simply put, there is a reduction in greenhouse gas emissions caused by the operation of public transportation itself, but it is also possible to increase the reliability of public transportation and convert from private cars and the like.

Next is "responding to more diverse needs." In our country, where public transportation as a mass transportation method during the period of high economic growth has once been established, it is required to be able to respond to various needs such as securing the first and last mile of movement that can be moved by public transportation door-to-door, securing personal space for enjoying the movement itself, and securing the means of unique movement during off-peak times.

The third point is "realizing more seamless movement." In modern times, it is common to move by changing various modes of public transportation. It is expected that new technologies will make movement beyond the barriers between modes and transportation operators more seamless.

The fourth point is "public transportation services that are accessible to everyone." Public transportation is a service that is literally open to the public, and users in various situations can be considered. In addition to infrastructure investments that lower physical hurdles such as barrier-free, the concept of equity that allows anyone to use it is also required in the consideration of the service level of public transportation.

4.4.3 Innovation in Vehicle Technology

4.4.3.1 Low Environmental Impact Vehicles

Reducing environmental impact in public transportation has become a mandatory condition. Of course, there are various measures for low environmental impact movement, such as getting people to use public transportation instead of private cars, but here we will organize the situation of low environmental impact vehicles as an innovation in vehicle technology.

4.4.3.2 Electric Buses

In the transportation industry, there is a great need for electrification of buses to reduce environmental impact. In foreign countries, various bus manufacturers are working on the development and sale of electric vehicles, and Japan is somewhat behind.

The introduction of electric vehicles to our country's route buses started with the introduction of the large bus "K9" made by BYD of China to the "Princess Line" connecting Kyoto City and Kyoto Women's University in 2015. In recent years, Iyo Tetsu Bus and Naha Bus have introduced vehicles from EV Motors Japan, which is based in Kitakyushu City and aims to assemble domestically (Photo 4.1).

Photo 4.1 EV bus by EV Motors Japan (Iyotetsu Bus in Matsuyama, Ehime). (Photo: Yurie Toyama)

4.4.3.3 Fuel Cell Buses

The fuel cell bus "SORA" that TOYOTA started selling in 2018 is being introduced to the Toei Bus and Tokyo BRT, among others, as an environmentally friendly bus.

4.4.3.4 Green Slow Mobility

According to the definition by the Ministry of Land, Infrastructure, Transport and Tourism [1–3], Green Slow Mobility is a small mobility service that utilizes electric vehicles that can travel at speeds less than 20 km per hour (Photo 4.2).

4.4.3.5 Hybrid Trains

In the railway industry, the operation of battery vehicles and diesel hybrid vehicles, and the research and development of railway vehicles using new power sources such as fuel cells are being promoted [4].

4.4.3.6 Miniaturization (Personal Mobility)

In order to meet diverse needs, the development of small vehicles is being promoted as a means of transportation for the last mile and as an alternative to private cars for relatively short distances. Many are used as sharing services, but in order to realize a sustainable service, it is necessary to install user-friendly ports, provide a

Photo 4.2 Green-slow-mobility service with vehicle type: e-COM-82, thinktogether (Miyazaki City). (Photo: Yurie Toyama)

sufficient number of vehicles to be available relatively anytime, and correct the bias of vehicles in one-way cases.

4.4.3.7 Micro Mobility

Small electric vehicles for one to two people are called "micro mobility," and from the early 2010s, demonstration experiments of sharing services were conducted in various places. Toyota Motor Corporation has launched a one-way car sharing service using Toyota Body's "COMS" in Toyota City, Chatan Town in Okinawa Prefecture, and Grenoble in France. Nissan Motor Co., Ltd., started the "Choimobi Yokohama" sharing service using the two-seater "Nissan New Mobility Concept" in the Minato Mirai area in 2013. However, all of these services have now ended their business and demonstration experiments (Photo 4.3).

4.4.3.8 E-Scooters

Various legal revisions are progressing, and the sharing service of E-scooters is spreading nationwide. On the other hand, in Paris, the sharing business of E-scooters was banned as a result of a citizen vote due to concerns about safety. There is a demand for the creation of mechanisms to make them safer and more convenient, along with rule-making.

Photo 4.3 One-way car sharing service with COMS (Grenoble, France). (Photo: Yurie Toyama)

4.4.3.9 Automation

There is a great expectation for autonomous driving from the perspectives of improving the safety of buses and addressing driver shortages. In our country, it has been set as a goal to realize "Level 4 automated driving mobility services" limited to certain areas in passenger services such as buses and taxis, at about 50 locations by fiscal 2025, and at more than 100 locations by 2027. Based on this government goal, demonstration experiments and discussions are being promoted in various places through projects such as the Ministry of Economy, Trade and Industry and the Ministry of Land, Infrastructure, Transport and Tourism's Smart Mobility Challenge and Road to the L4.

The Ministry of Land, Infrastructure, Transport and Tourism has been working on demonstration experiments of autonomous driving services based at roadside stations in mountainous areas since fiscal 2017. Vehicles such as small buses and modified golf carts were used, and demonstration experiments of autonomous driving services were conducted as a means of transportation for local residents and tourists, with roadside stations as the base. In 2019, the operation system was transferred to the local area in Kamikoani Village, Akita Prefecture; and Higashiomi City, Shiga Prefecture, and the phase of "social implementation" began.

In addition, BOLDLY has provided services using vehicles from the French company NAVYA in Sakaimachi, Ibaraki Prefecture; JR West and Softbank are working on demonstration experiments of platooning of buses, and demonstration experiments and social implementation are progressing nationwide (Photo 4.4).

Photo 4.4 Automated bus service (Sakai Town, Ibaraki). (Photo: Yurie Toyama)

4.4.4 Innovation in User Experience

Technological innovation in public transportation is not just about vehicle technology. For example, mobile tickets that are highly compatible with the concept of Mobility as a Service, and technological innovations such as trip planners (transfer search), have made public transportation more user-friendly and convenient. Here, these technologies are organized as technological innovations related to user experience.

4.4.4.1 Real-Time Dispatch

Looking back in history, a dispatch service called "Dial-a-Ride" using the telephone was introduced in Sweden in the 1970s. Since then, with the innovation of various ICT technologies, the improvement of communication speed, and the evolution of computer systems that can calculate complex algorithms, it has become possible to receive requests, calculate the optimal route, and dispatch vehicles in real time.

4.4.4.2 Cashless Payment

If users can ride public transportation using the smartphones and credit cards they always use, it would be convenient not only for users who do not use public transportation on a daily basis, but also for tourists and others who only use public transportation in the area for a limited period of time. In addition, digital ticketing on smartphones can adopt more flexible and diverse fare services and discount fare application conditions than paper planning tickets, and it is also a great benefit to public transportation operators in terms of easy acquisition of data such as usage history.

4.4.5 Dealing with New Technology

The new technology surrounding public transportation is evolving daily. In order to make the use of new technology effective and sustainable, it is important not to make the introduction of new technology a goal, monitor the effects and the acceptance of the use of new technology by the transportation operators themselves and the users on a long-term and regular basis. Also, boldly rearranging operations and designs that should be reviewed due to the use of new technology can have an impact on the region and lead to an improvement in the attractiveness of public transportation.

When breaking down the elements of so-called new technology, the digitization of data, the dramatic improvement in computer functions, the ability to perform large-scale and high-speed calculations and store large amounts of data, and the dramatic improvement in communication technology, enabling large-scale and high-speed data communication, are the starting points. On top of that, the internet, artificial intelligence, etc., have become the foundation, and various services have become possible. From a broad perspective, visualization and efficiency can be mentioned. Various phenomena can be visualized, and various administrative processes can be realized at low cost and high speed.

From the standpoint of public transportation policy and regional policy, the points of particular attention are said to be visioning and validation. If I were to interpret it in my own way, in scenes where various things can be visualized with new technology and various tasks are being made efficient, we are being asked to have a clear vision of the future regional society and the public transportation that will support it. And there, various hypotheses will be set. Some hypotheses are verified in the computer, in a virtual space, or in the metaverse, or in the digital twin. Some hypotheses are verified in the form of demonstration experiments or social experiments in the actual regional space, based on an understanding of the region. In either case, diverse, large, and complex data will be obtained, and the ability to verify using them will be required.

While more and more young people are learning data science, it is necessary not only to learn simple tool usage and programming methods, but also to learn the philosophy underlying them, and the philosophy of visioning and validation in the real world where they are positioned.

4.5 Public Transportation and Future Society

In the process of re-examining public transportation on the roads that support local communities and promoting necessary reconstruction (re-design) for the future, the following keywords can be referenced. First, as basic terms, we have multi-modal and freedom of travel. Next, for specific scenarios, we have walkable, reliable (and trust), and enjoyable. These will be explained in order.

4.5.1 Multi-modal

In Japan, this term is sometimes used in the context of public transportation connections, but in its original meaning, it refers to scenarios where multiple modes, or means of transportation, can be chosen. It's not a situation where there are no means of transportation other than driving a private car, but rather, even if you can't drive a private car, you can move by other means, often by public transportation. In this sense, it is necessary to prepare a multi-modal transportation environment. In recent years, the well-being indicator is often used in the evaluation of smart city policies. The premise of well-being is subjective happiness, which can be experienced in an environment where choices can be made. When you want to go out, having multiple destinations to choose from, being able to choose not only private cars but also other means such as public transportation to get there, and being able to choose both fast and slow methods of travel, being able to travel quietly alone or lively with friends, etc., having multiple options greatly contributes to subjective happiness. Therefore, in transportation planning, it is important to provide a transportation service worth choosing and to make people aware of it.

4.5.2 Freedom of Travel

Among derived demands, there is a variety from movements with a strong sense of obligation to those without. Independently of this, there is also a variety from movements with a strong sense of excitement and lingering satisfaction to those completely devoid of such feelings. Among these, even if there is a slight sense of obligation, when you can enjoy the excitement before going out, the activities at the place, and the pleasant afterglow after finishing, for example, a play, you feel glad that you went out and can feel the freedom of movement. Going back to the basics, even if there is some waiting time or cost burden, if you want to go out, it can become a reality, and if you want to take a detour, it can come true. You can feel the freedom of movement in such things. In the case of wheelchair users boarding a bus, aside from cases where boarding is outright refused, even if they can board, if they are firmly fixed to the handrails or floor by someone else's strength, the wheelchair user themselves completely loses their freedom. Despite such situations being easy to imagine, for example, in Japan's buses, such situations are not improving at all. Even if the bus vehicle becomes low-floor, unless such regulations and the structure of the route to the bus stop and the bus stop part are improved, wheelchair users will not be happy to board the bus.

4.5.3 Walkable

In recent years, the phrase "creating a city enjoyable to walk in" has become well-known, and the word "walkable" has also become commonly used. In its original meaning, "walkable" implies that it is possible to walk or reach a destination by walking.

In terms of public transportation on the roads, it is important to be able to walk safely and comfortably from one's home or other starting point to a bus stop or tram stop, and from the bus stop or tram stop to the final destination. It is also important to be able to wait safely and comfortably at the bus stop or tram stop.

While some may think that door-to-door public transportation, such as the so-called DRT (demand responsive transport (transit)), is the standard for the future, it

4 Public Transportation and the Local Community

Photo 4.5 Pinhais bus terminal, Curitiba, Brazil. (Photo: Fumihiko Nakamura)

is not very realistic to expect complete door-to-door service at all times and in all areas considering the efficiency of moving vehicles and the maintenance of people's health. Therefore, bus stops and tram stops, which may vary in placement and density, are very important gateways. Ideally, the cost of these should be borne by the road administrators, and it is important that it is possible and comfortable to walk to these stops (Photo 4.5).

4.5.4 Reliable and Trust

For mainline public transportation, including buses, trams, and even LRTs and BRTs, punctuality and speed are often expected together. There are many examples of setting KPIs such as whether the service is running on schedule and whether the required time is stable, and attempting to evaluate them by running the so-called PDCA cycle.

However, for local residents and stakeholders, the issues are not there. First, it is important to know about the service; understand its characteristics, strengths, and weaknesses; and trust and accept it.

Even if there are only a few services a day, and even if they are always 15 minutes late on rainy days, trust and reliability can be born when these facts are shared and accepted.

There are many cases where buses are labeled as being late or unusable based on impressions, feelings, and word of mouth, and are not recognized at all. In the age of data, it is expected that the operation will always be visible and that this will increase trust (Photo 4.6).

Photo 4.6 Kiyohara transit center, Utsunomiya Light Line, Tochigi, Japan. (Photo: Fumihiko Nakamura)

4.5.5 Enjoyable

In the comparison between private cars and public transportation, private cars are often portrayed as luxurious and comfortable vehicles, while public transportation is seen as a vehicle to be used with patience. Traveling by public transportation is seen as a hardship and a test of patience. It is often thought that if it is cheap, it can't be helped if it is bad.

Instead, it is hoped that the idea of actively creating situations where public transportation is more enjoyable, not just leaving it up to the operators, will be embraced. There are many ideas, such as having a public market in front of the bus stop or showing short films inside the vehicle (Photo 4.7).

4.6 Conclusion

In this chapter, we have summarized the basic perspectives and issues related to public transportation, which is rooted in local communities and contributes to the realization of a safe and comfortable sustainable society, as well as related technological trends. We have not touched on recent discussions about ride-sharing and autonomous driving, but we have organized the issues of basic demand structure, regional issues, and technological issues.

In order to sustain local communities, providing means of transportation other than private cars also contributes to reducing the accident risk associated with the driving of elderly people who have issues with their driving skills. Including the example of children, it goes without saying that public transportation rooted in local communities is also linked to the realization of a sustainable future and a society without traffic accidents.

Photo 4.7 Cafe with Pacific Ocean view at Hitachi Railway Station, JR East, Japan. (Photo: Fumihiko Nakamura)

References

1. Niitani Y (ed) (2003) Urban transportation planning, 2nd edn
2. Brian Richards, Transport in cities (1992)
3. Ministry of Land, Infrastructure, Transport and Tourism, Green Slow Mobility., https://www.mlit.go.jp/sogoseisaku/environment/sosei_environment_fr_000139.html, Viewed on October 30, 2023
4. Ministry of Land, Infrastructure, Transport and Tourism, Study Group on Accelerating Carbon Neutrality in the Railway Sector, https://www.mlit.go.jp/tetudo/content/001611770.pdf, Viewed on October 30, 2023

Open Access This chapter is licensed under the terms of the Creative Commons Attribution-NonCommercial-NoDerivatives 4.0 International License (http://creativecommons.org/licenses/by-nc-nd/4.0/), which permits any noncommercial use, sharing, distribution and reproduction in any medium or format, as long as you give appropriate credit to the original author(s) and the source, provide a link to the Creative Commons license and indicate if you modified the licensed material. You do not have permission under this license to share adapted material derived from this chapter or parts of it.

The images or other third party material in this chapter are included in the chapter's Creative Commons license, unless indicated otherwise in a credit line to the material. If material is not included in the chapter's Creative Commons license and your intended use is not permitted by statutory regulation or exceeds the permitted use, you will need to obtain permission directly from the copyright holder.

Chapter 5
Traffic Engineering

Takashi Oguchi, Miho Iryo, and Azusa Toriumi

Traffic engineering is a field of science and technology aimed at achieving safe, smooth, and sustainable movement of people and goods. Traffic engineering emerged in the early twentieth century with the spread of internal combustion engine vehicles and developed with the global motorization after World War II. Its initial scope was planning, design, and operation of roads necessary to reduce traffic congestion and accidents caused by automobiles. Now, it deals with a wider range of traffic issues, considering sustainability of people's mobility, including not only safety and efficiency but also traffic pollution (environmental load) for not only automobiles but also pedestrians, bicycles, and public transportation (such as buses, trams, and LRT). This chapter provides an overview of the techniques to realize smooth and safe road traffic and the methodologies to scientifically analyze and describe the flow caused by the movement of multiple road users (i.e., traffic flow), as the basics of traffic engineering.

5.1 Fundamentals of Road Traffic

5.1.1 Description of Traffic Flow

First, we will define the variables to describe the flow of traffic. The flow of automobile traffic is represented by a time-space diagram as shown in Fig. 5.1. This is a diagram that takes time on the horizontal axis and distance from a reference point on the vertical axis and represents the trajectory of each vehicle. Here, the time

---

T. Oguchi (✉) · A. Toriumi
Institute of Industrial Science, The University of Tokyo, Meguro, Japan
e-mail: takog@iis.u-tokyo.ac.jp

M. Iryo
Graduate School of Environmental Studies, Nagoya University, Nagoya, Japan

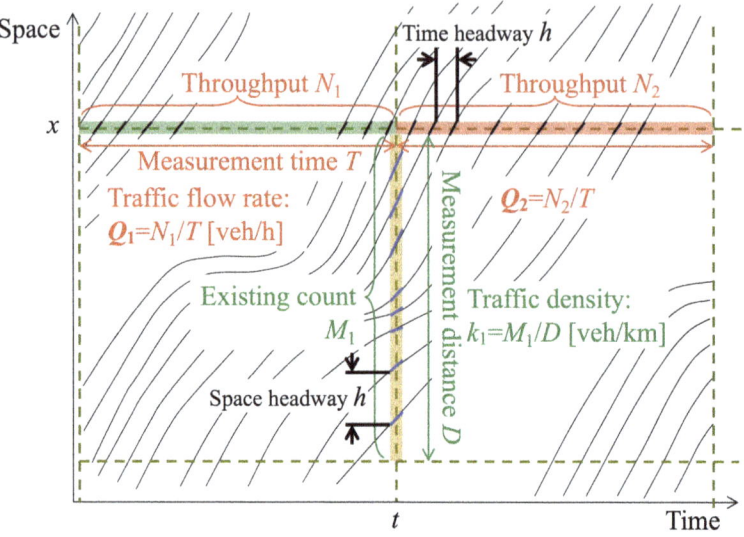

Fig. 5.1 Time-space diagram

interval from when the front of a vehicle passes a certain point until the front of the next vehicle passes is called the time headway, and the time from when the rear end of a vehicle passes a certain point until the front of the next vehicle passes is called the gap time. Also, at a certain time, the distance from the front of the previous vehicle to the front of the following vehicle is called the distance headway, and the distance from the rear end of the previous vehicle to the front of the following vehicle is called the gap distance.

Next, using this time-space diagram, we define the traffic volume, traffic density, and speed that represent the flow of traffic. The number of vehicles passing a certain point in 1 h or less, or this number per unit time, is called the traffic flow rate [vehicles/hour] (generally also called traffic volume). Traffic density is the number of vehicles present at a certain point in time, converted to the number of vehicles per unit distance [vehicles/km], with 1 km or less in the direction of road extension as the aggregation unit.

The speed [km/h] is a representative speed of traffic flow consisting of individual vehicles with different speeds. This representative value uses the "space mean speed." The space mean speed is defined as the average speed of vehicles existing in a unit road section at a certain time, but it is difficult to directly observe this. If the vehicle speed and headway in the traffic flow are in a stable state where they do not change (steady state), the space mean speed can be estimated as the harmonic mean of the vehicle speeds observed at a certain cross section.

5.1.2 Basics of Traffic Flow

Between the traffic volume Q, traffic density K, and (space mean) speed V, there is a relationship as,

$$Q = KV \tag{5.1}$$

Note that Eq. 5.1 generally does not hold if the speed values of each vehicle observed at the point are simply (arithmetically) averaged. Equation 5.1 corresponds to the basic law of physics, the "law of mass conservation," and universally holds for "flows in general," not limited to traffic.

Figure 5.2 shows the mutual relationship of these three variables using Eq. 5.1. The traffic volume and speed were measured and aggregated every 5 min by vehicle detectors installed on the road. As the traffic density approaches zero, the speed reaches its maximum value (free speed), and as the speed approaches zero, the traffic density reaches its maximum value (jam density), both of which are roughly monotonically decreasing. Also, there exists a traffic density (critical density) and speed (critical speed) where the traffic volume reaches its maximum value. A state with a higher density than the critical density (with a speed lower than the critical

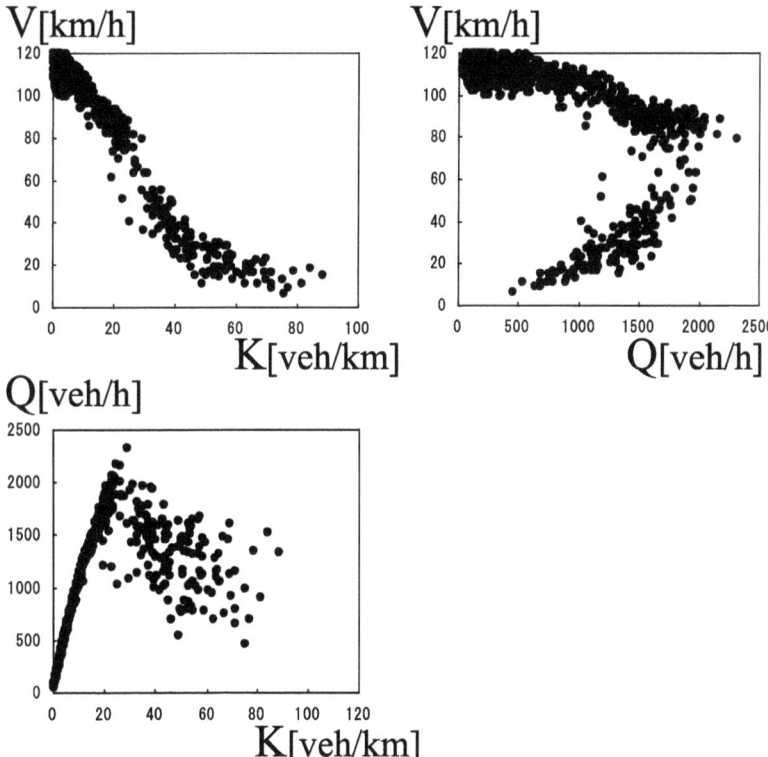

Fig. 5.2 Measured examples of traffic volume, traffic density, traffic speed relationships

speed) is a traffic congestion state, and conversely, a state with a lower density than the critical density (with a speed higher than the critical speed) represents a non-congested traffic state. The state around the critical density is sometimes called the critical flow state.

As shown in Fig. 5.2, there is generally a monotonically decreasing relationship between traffic density and speed. Combining this with the relationship of Eq. 5.1, the traffic volume-traffic density (traffic volume-speed) relationship is a two-value function, and the existence of the maximum value of traffic volume is derived. This is the theoretical basis for the existence of the "maximum traffic volume (traffic capacity)" in road traffic.

Until now, we have dealt with the flow of vehicles moving in one direction along a lane. However, in pedestrian traffic, the concept of lanes does not exist, and people can move in various directions. Also, in developing countries, vehicles may not follow lanes and may run side by side in numbers greater than the number of lanes. Even in such non-lane-following traffic, for a steady (no bias in speed or density) flow moving in one direction along straight, elongated sections such as roads or passages, traffic flow can be handled in the same way as above by defining traffic volume as the number of vehicles/pedestrians passing through a certain cross-section per unit time and width, and traffic density as the number of vehicles/pedestrians per unit area within the section. In pedestrian traffic, in places like open spaces where people can move in all directions, or at corners, there may be cases where it is difficult to divide the space into appropriate units because pedestrians concentrate at specific points. In such cases, some methods determine the area of space that each person occupies based on relative positions between adjacent people.

5.1.3 Traffic Congestion

The traffic capacity of a road is determined by various factors such as the number of lanes, changes in gradient or changes in light and dark at the entrance of tunnels [1, 2], and restrictions on departing time due to traffic signals, so the traffic capacity varies for each road section. Locations with relatively low traffic capacity compared to adjacent sections are called bottlenecks in terms of traffic capacity. Traffic congestion is defined in traffic engineering as "a state where a queue of vehicles remains on the upstream side of the bottleneck when traffic demand exceeding the traffic capacity of the bottleneck arrives." The queue of vehicles itself may also be referred to as traffic congestion. On Japanese highways, the sag of the road (where the longitudinal gradient changes from downhill to uphill) and the entrance to tunnels are the main bottlenecks, accounting for about 80% of all congestion locations, while in Europe and the US, merging and weaving sections are the main bottlenecks. On surface streets, signalized critical intersections with high traffic demand, queues waiting for parking at large shopping centers, and sections with on-street parking are likely to become bottlenecks. In addition to congestion caused by traffic concentration at specific locations on the road, there are also sudden traffic congestoins caused by traffic crashes that temporarily block some lanes and become bottlenecks, and traffic congestions caused by lane closures due to road construction. Based on the above definition of traffic congestion, it is not possible to determine traffic congestion simply by a threshold of traffic speed. On Japanese intercity expressways, speeds below 40 km/h, and on metropolitan expressways, speeds below 20 km/h are used as criteria for determining traffic congestion, but these are merely practical and convenient thresholds. When traffic demand exceeding the traffic capacity of a bottleneck arrives, traffic congestion forms upstream section of the bottleneck. At the upstream end of the congestion, the traffic flow discontinuously transitions from the upstream non-congested flow to the downstream congested flow, and the position of the upstream end of the congestion extends backward at a certain speed. The extension speed of the upstream end of the congestion can be calculated using shock wave theory [3]. Assuming that the relationship in Fig. 5.2 can be simplified as in Fig. 5.3, and the traffic volume-traffic density relationship can be schematically represented by a triangular function. The traffic state of traffic demand A arriving at the bottleneck C_b (Point BN in Fig. 5.4) is represented by Point a (with traffic volume A, traffic density k_1, and speed v) in Fig. 5.3. Here, if $A>C_b$, traffic congestion occurs in the upstream section of the bottleneck, and its state is represented by Point b (with traffic volume C_b, traffic density k_2, and speed v_b) in Fig. 5.3, and the upstream end of the congestion moves backward. The speed u_{ab} of the discontinuous front between these traffic states (shock wave) is given by Eq. 5.2.

$$u_{ab} = (A - C_b) / (k_1 - k_2) \quad (5.2)$$

Equation 5.2 means the slope connecting Points a and b in Fig. 5.3, and u_{ab} in the figure has a negative slope, so the upstream end of the traffic congestion moves backward. In Fig. 5.4, the boundary between the non-congested state a and the

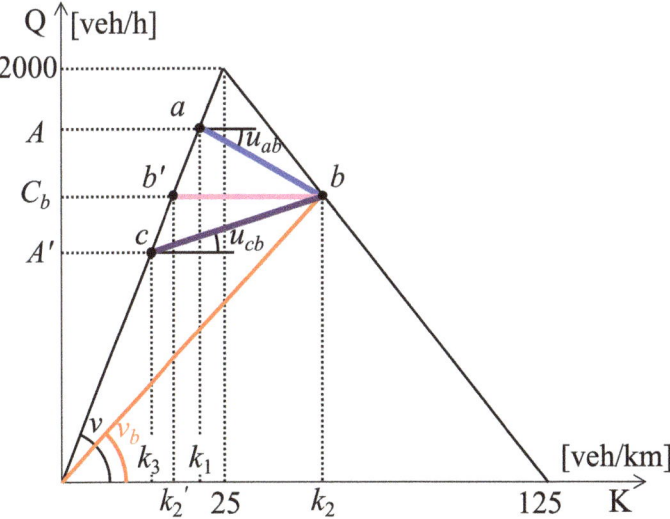

Fig. 5.3 Traffic volume-density relationship model

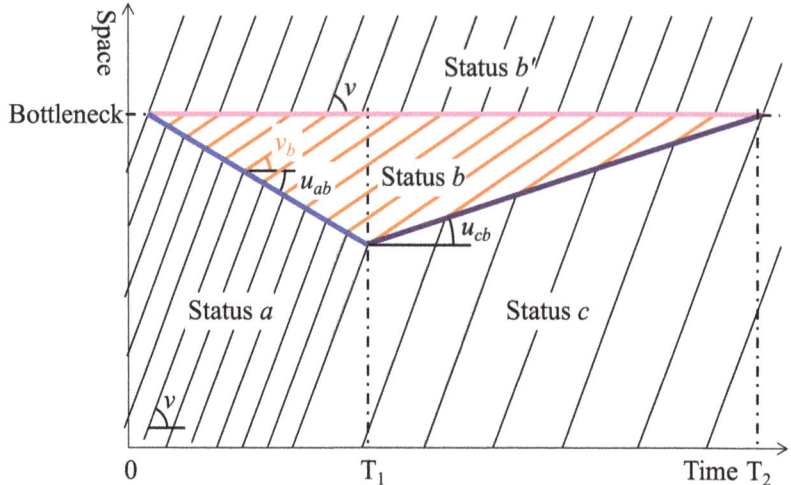

Fig. 5.4 Shockwave of traffic congestion extension and reduction

congested state b moves at speed u_{ab}. Here, if the traffic demand changes to point c (with traffic volume A', traffic density k_3 and speed v) in Fig. 5.3 at time T_1, and $A' < C_b$ holds, the slope u_{cb} connecting Points b and c is positive, so the upstream end of the traffic congestion, which is the boundary between b and c, shrinks and disappears at time T_2.

For example, when traffic demand $A = 1800$ [vehicles/hour] arrives at the bottleneck traffic capacity $C_b = 1600$ [vehicles/hour], $u_{ab} = (1800-1600)/(22.5-50) \fallingdotseq -7$ [km/h] indicates the extension of traffic congestion, and if $T_1 = 2$ hours, then the maximum length of the traffic congestion, L, is approximately 14 km. If the traffic demand decreases to $A' = 1500$ [vehicles/hour], then $u_{cb} = (1500-1600)/(18.75-50) = 3.2$ [km/h], and the time T_2 is $T_1 + L/u_{cb} \fallingdotseq 6$ h and 30 min.

Generally, even if traffic congestion occurs due to an increase in traffic demand, the excess percentage of traffic demand over the traffic capacity of a bottleneck is at most several percent (in this calculation example, $1800/1600 = 1.125$, so the excess percentage is 12.5%). It is known that the duration of traffic congestion (6.5 h in this case) is considerably longer than the time when traffic demand exceeds (2 h in this case) [4].

However, on actual roads, the traffic capacity at a single bottleneck is not constant. At sags and tunnel entrances, which are the main bottlenecks on Japanese expressways, slight speed disturbances due to gradient changes can trigger deceleration waves that propagate backward and change into traffic congestion. The traffic demand that causes traffic congestion varies greatly within the range of 75–90% of the normal traffic capacity of a simple road section. This is thought to be due to individual and vehicle differences in the car-following behavior of drivers [5].

Once traffic congestion occurs, the traffic capacity of a bottlenck further decreases, dropping to around 60% of the normal capacity of a simple section. This is due to the fact that boredom and fatigue occur in the slow traffic congestion state, and the following behavior becomes sluggish. In the above example, the traffic capacity was set at a constant value of 1600 vehicles/hour, but if this value is reduced, it is found that the propagation speed and duration of traffic congestion increase.

5.1.4 Road Network Traffic Flow

As explained in the previous section, traffic congestion is perceived as a local phenomenon occurring around bottlenecks, and its observation is made for each individual road section. On the other hand, when focusing on the entire road network of a city, bottlenecks are scattered everywhere, and it is required to grasp their state in a planar manner. Traffic congestion is a nonlinear phenomenon that occurs when traffic demand exceeds the traffic capacity of a bottleneck, and once congestion occurs, users choose routes to avoid it, so it is not enough to focus on a single bottleneck on the network, and a planar evaluation is necessary.

As a macroscopic indicator that captures the overall traffic situation of the entire road network in an aggregated manner, the Macroscopic Fundamental Diagram (MFD) is used. The MFD shows the relationship between the spatial average density (or number of vehicles present) and the spatial average traffic volume (or trip completion flow rate, also known as throughput) on a road network. An example of this is shown in Fig. 5.5. Similar to the traffic volume-density relationship for an individual road section in Fig. 5.2, a state where the throughput is maximized at a certain spatial average density can be observed, and if the density increases beyond that, the throughput decreases [6]. When throughput is at its maximum, it can be

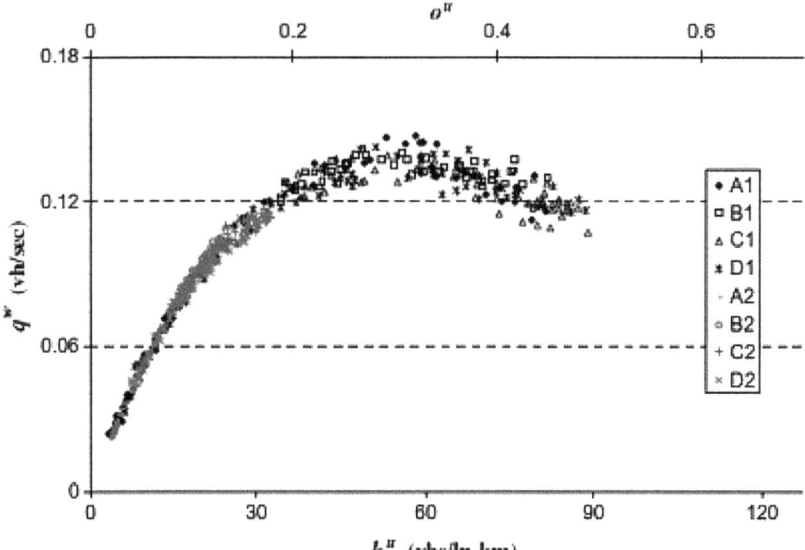

Fig. 5.5 Example of macroscopic fundamental diagram [6]

interpreted that the performance of the entire network is maximized. If the spatial average density increases beyond that condition, the overall network performance decreases due to the effects of traffic congestions scattered in the network. Therefore, the MFD is expected to be used as an evaluation method for controlling the inflow to the expressway networks, etc., and research on understanding the characteristics of the MFD and control methods has been advancing in recent years.

5.2 Road Traffic Safety

5.2.1 Traffic Crash Occurrence

Among the statistical information on road traffic crashes, the number of fatalities in Japan is usually described as the number of 24-h fatalities (the number of people who died within 24 h after the crash) in police statistics. The number of fatalities generally used by the World Health Organization (WHO) is the number of people who died within 30 days of the crash, and the number of fatalities within 30 days is used when conducting international comparisons. It should be noted that crashes that are not reported cannot be recorded as statistical data, so it is not possible to fully grasp the actual situation of minor injury crashes and others depending on the country and type of crashes.

According to the Global status report issued by the WHO in 2018 [7], the number of traffic crash fatalities worldwide is about 1.3 million per year, making it the leading cause of death among young people aged 5–29. In addition, it is estimated that 20–50 million people or more are injured in traffic crashes each year. More than half of the crash fatalities are vulnerable road users such as pedestrians, cyclists, and motorcyclists, and 93% of traffic crash fatalities occur in low- and middle-income countries.

The number of traffic crash fatalities in Japan (24-h fatalities) recorded the highest number of fatalities at 16,765 in 1970. After reaching a second peak in 1992, it has been decreasing, and in 2022 it was 2,610. The transition of the number of crash fatalities per 1 million population (30-day fatalities) in recent years is compared internationally in Fig. 5.6. Except for the United States, there is a general trend of decrease. In addition, the state-specific composition ratio and age-specific composition ratio of traffic crash fatalities are compared internationally in Figs. 5.7 and 5.8, respectively. Compared to other OECD countries, Japan is characterized by a very high number of fatalities among pedestrians and the elderly.

Looking at the situation of traffic crashes in Japan by road types (2022), crashes at intersections and near intersections account for more than half, at 57%. In terms of crash types, rear-end collisions are the most common at about 30%, followed by head-on collisions at about 25%.

5 Traffic Engineering

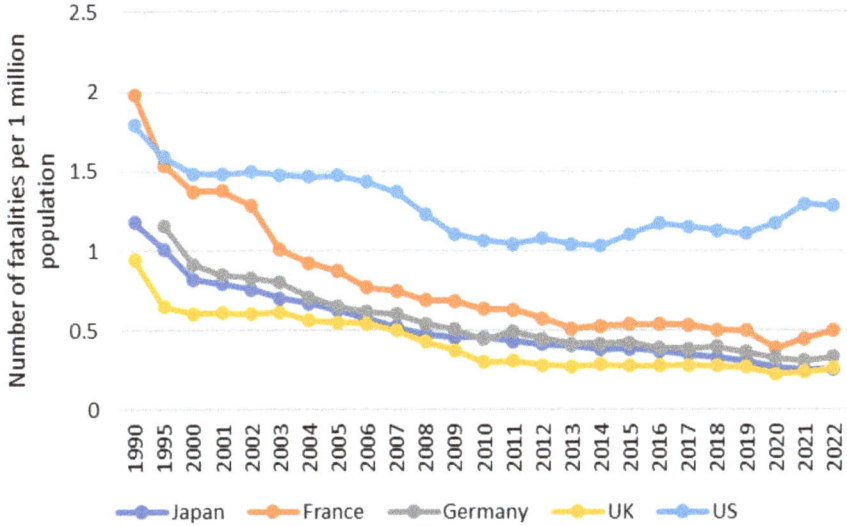

Fig. 5.6 International comparison of number of fatalities per 1 million population (from IRTAD)

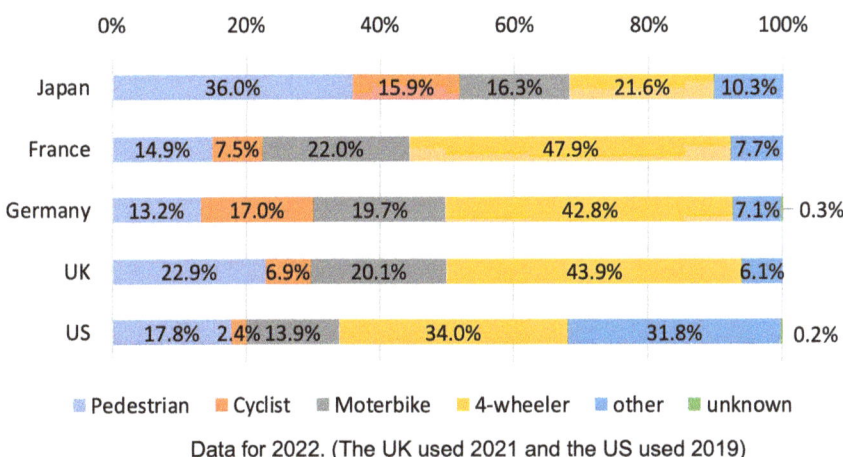

Data for 2022. (The UK used 2021 and the US used 2019)

Fig. 5.7 International comparison of crash composition ratio by road user type (from IRTAD)

5.2.2 *Crash Risk Assessment*

To consider countermeasures against traffic crashes, it is necessary to properly understand the risks involved in traffic crashes. Various factors, such as road geometry, traffic conditions, land use, and driver characteristics, affect the occurrence of traffic crashes.

Among various factors, focusing on traffic flow, it has been shown that in rear-end collisions, the crash rate (number of crashes per unit traffic volume) increases

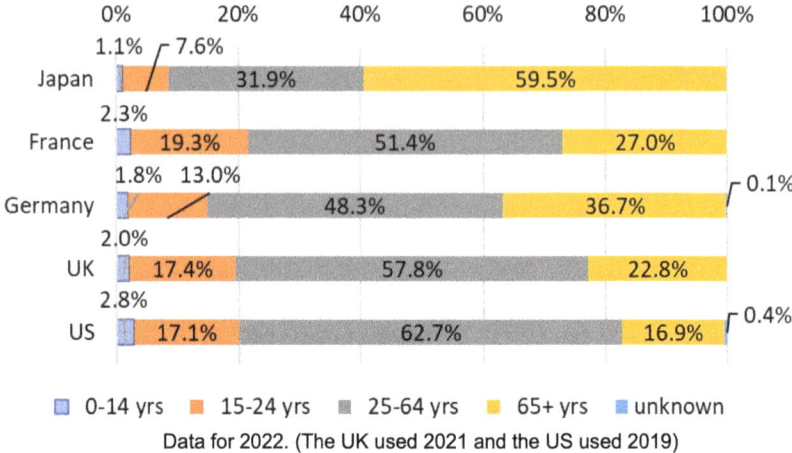

Fig. 5.8 International comparison of crash composition ratio by age (from IRTAD)

as the traffic volume increases, and conversely, the crash rate of single-vehicle crashes decreases [8]. When traffic volume is high, crashes can occur due to a following vehicle colliding at the upstream end of a traffic congestion. Also, in situations where free flow and traffic congestions are mixed temporally and spatially, it is known that crashes such as vehicle-to-vehicle collisions during lane changes occur due to differences in vehicle speeds.

Also, it is known that traffic crashes frequently occur at specific points. For example, on main roads, about 60% of fatal and injury crashes are concentrated in 12% of the sections (in 2015–2018), so it is effective to take concentrated measures against crash-prone locations.

Although traffic crash analysis is basically a statistical analysis using actual traffic crash data, traffic crashes are rare events, so it takes several years to collect statistically significant data. As a method to understand the danger level of the target point in a short time or to measure the effect of countermeasures, there is an evaluation method using Surrogate Safety Measure (SSM). SSM is an indicator to evaluate potential collision danger from the behavior of road users. Various indicators are proposed depending on the types of conflicts, and typical ones include the passing time difference at the conflict point when the trajectories of the two intersect (Post Encroachment Time: PET), and the time until collision when two vehicles moving in the same direction continue to maintain their current speed (Time to Collision: TTC). In recent years, it has also been conducted to extract the frequency of sudden behaviors such as sudden deceleration from probe data such as ETC2.0 (refer to Sect. 5.5.1) and evaluate it.

5.3 Traffic Function and Planning, Design, and Operation of Roads

This section provides an overview of the basics of planning, design, and operation of roads.

5.3.1 Functions and Hierarchical Network of Roads

Functions of roads can be divided into traffic function and spatial function. Among these, the traffic function is for the movement of people and goods, which is primary required on roads. The traffic function is further categorized into three: through-movement function (also called movement function or mobility function), land-access function (also called access function), and staying function. These can be defined for different modes of transportation such as automobiles, pedestrians, and bicycles. The through-movement function represents the ability of road users to smoothly pass through the road, the land-access function represents the ease of entry and exit from the road to roadside facilities and other roads, and the staying function represents the ability of road users to stay on the road (e.g., automobiles' loading/unloading and on-road parking, people's pausing). On the other hand, the spatial function is enabled by the space created by the existence of the road. In particular, urban roads have spatial functions such as urban skeleton formation, promotion of roadside development, disaster prevention by preventing fire spread, greening and landscape, and accommodation of lifelines (e.g., electric wire, water, and sewage) and transportation facilities (e.g., subway, underground parking).

The through-movement function, which is the primary function of roads in the traffic function, is in conflict with other two functions, namely, the land-access and staying functions. In other words, land-access and egress or on-road staying can be a hindrance to users trying to conduct through-movement, and vice versa. Therefore, a single road section cannot bear all of these functions, and it is efficient to assign priority of these functions to each road while achieving all the necessary functions as a whole road network. This classification of individual roads according to the priority of through-movement, land-access, and staying functions create hierarchical organization of a road network, thus is called a "functionally hierarchical road network."

The hierarchical organization of road network allows smooth and high-speed movement on roads where automobiles' through-movement function is prioritized by avoiding land-access and staying. This typical example includes motorways and major arterial roads. Conversely, on roads where through-movement function is restricted, high-speed automobiles are eliminated, making it easier for land-access and staying, or creating spaces for pedestrians and bicycles. Such examples include main streets in CBDs and residential alleys. There are various scales of movement, from daily short-distance movement to long-distance intercity movement. The

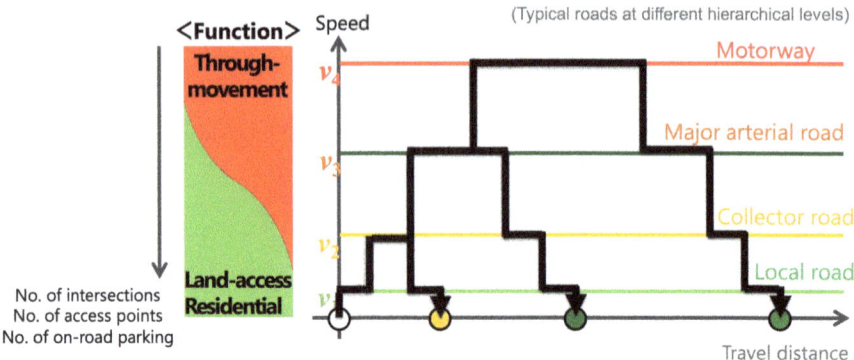

Fig. 5.9 Conceptual relationships between road hierarchy and movement scale

functional hierarchical road network allows for the use of roads at different hierarchical levels according to such movement scales, such as motorways for long-distance movement and residential alleys for short-distance movement, thereby achieving overall efficiency (see Fig. 5.9). Separating road users who interact with each other into roads at different hierarchical levels is expected to improve safety. By placing roads with priority on through-movement functions outside residential areas and city centers, where through traffic can be eliminated, thus improving living and urban environments, such as noise and exhaust gas reduction.

5.3.2 Quality of Service and Road Performance Evaluation

In road planning, designing, and operation, it is important to clarify the functions that each road section should fulfill and ensure to achieve them. In other words, at each stage of planning, design, and operation, it is important to maintain the expected function by evaluating the quality of service (QoS) perceived by road users using various service measures. Service measures, for example, for automobile through-movement function include traffic density, indicating the number of vehicles occupying the road space, delay, which is the extra time required for waiting at signalized intersection and so on, travel time or travel speed, which is measured at the certain section including delay, and the proportion of vehicles forced to follow the leading vehicle without being able to overtake. Similarly, for pedestrian through-movement function, service measures include not only the degree of congestion in walking space and travel time, but also perceived safety and security under the presence/absence of sidewalks and exposure to cars and bicycles. The evaluation targets of service measures can be set in various scales, from individual intersections, road sections with several intersections, routes consisting of multiple road sections, to the entire road network of a region, corresponding to the subject of planning, design, and operation.

Research on evaluating the QoS of roads has been developed firstly in the United States, where motorization progressed faster than others, and these results have been summarized and regularly published as the guidelines, the Highway Capacity Manual (first edition 1950–latest 7th edition 2022) [9]. Since the second edition published in 1965, a concept has been proposed to evaluate the QoS using the level of service (LoS), which classifies the qualitative satisfaction of service measures from A to F. This concept has continued to evolve, expanding its scope from automobiles to multimodal means of transport including pedestrians, bicycles, public transportation, and trucks. In Japan, research and practical deployment of "performance-oriented" road planning, design, and operation are progressing [10, 11]. In such a framework, the target performance is set depending on the QoS required for roads at different hierarchical levels then evaluated whether achieved or not. Traditionally, these studies have mainly focused on automobile traffic. However, in recent years, as attention is being paid to more sustainable means of transport because of various reasons such as environmental issues, revitalization of cities, and health promotion, it is required to consider diverse users such as pedestrians, bicycles, public transportation, and transport for logistics.

The QoS and service measures realized on roads are influenced not only by the conditions of geometry and traffic control of the roads but also by dynamic conditions such as traffic volume as outlined in Sect. 5.1. Therefore, even on the same road, service measure is not always constant, but changes from moment to moment and day by day. The average during the target time is often used as a representative value for evaluation, but the consistency or dependability of the service measure against the daily or hourly fluctuation is also worth evaluated as "reliability." For example, travel time reliability is very important for people to decide their departure time in order to arrive at their destination on time based on their part travel experiences. As an index to measure travel time reliability, percentile values obtained from the distribution of travel times measured within the target time period are often used. Furthermore, evaluation of connectivity reliability and redundancy of road network is also important considering disasters.

5.3.3 *Planning, Design, and Control of Intersections*

Intersections where roads intersect are one of the most critical points of a road network that affect both traffic safety and efficiency. If roads intersect at grade, traffic flows that intend to go straight are disturbed by other traffic flows to some extent, which deteriorating through-movement function. Therefore, when planning, designing, and operating intersections, it is important to pay close attention to maintain the functional hierarchy of the road network.

Expressways, where automobiles' through-movement function is most prioritized, minimize intersections by limiting access-egress function to other roads, and apply grade-separation (interchange) where connection is necessary so that traffic flow is not disturbed. Even on surface streets, major arterial roads, which should serve for a certain level of through-movement function, generally require some geometric treatments, traffic control, and regulations to avoid frequent disturbances of through traffic flow at the intersections with minor roads, as illustrated in Fig. 5.10.

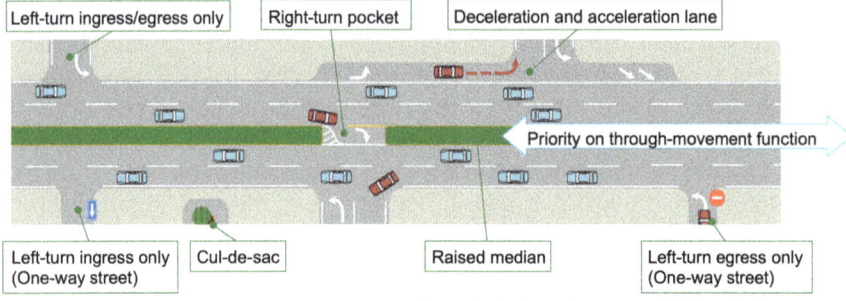

Fig. 5.10 Schematic of access management

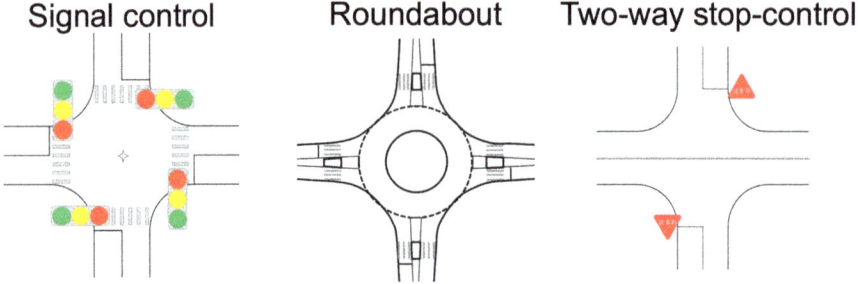

Fig. 5.11 Examples of traffic controls for at-grade intersections

This idea is systematically considered as "Access Management" [12] in the United States. Moreover, it is also important to reduce delay at intersections where major arterial roads cross each other by partially elevating some directions of the roads.

When making intersection at grade, it is necessary to select an appropriate type of control according to the traffic demands of different road users for directions as well as the surrounding environment. Signal control is one of the most typical controls for at-grade intersections, but not necessarily be the most appropriate because it incurs delay even under extremely low traffic demand. Studies have been conducted to develop methodologies for evaluating traffic efficiency (such as delay) and safety of intersections under different types of intersection controls, such as signal control, roundabout, and unsignalized control (i.e., two-way / all-way stop-control) (Fig. 5.11) according to traffic demands and other conditions.

In many countries, pedestrian safety is a major problem. Typically, there are problems that automobiles do not yield to crossing pedestrians, and pedestrians ignore crosswalks and traffic lights (jaywalking). With the establishment of a functionally hierarchical road network, it is fundamentally important to minimize/reduce automobile traffic in places where many pedestrians are expected. In addition, it is crucial to provide appropriate crossing opportunities for pedestrians where demanded. Consideration must be made to determine the location of crosswalks (midblock or intersection), presence or absence of traffic lights, and its geometry depending on the conditions.

5.3.4 Traffic Signal Control

Signal control is required for at-grade intersections with relatively high traffic volume. At signalized intersections, time, right of way in other words, is allocated for various directions of traffic flow. "Stage" (also named "Phase") can be defined as the right of way given simultaneously to a set of traffic flows (including automobiles, pedestrians, and/or bicycles), or the time duration in which the right of way is allocated. A series of stages applied for all traffic movements to cross an intersection is called a "signal program" (also called "signal plans"). One of the most typical and simplest signal programs is a two-stage control, which alternately allocate green times to intersecting roads as shown in Fig. 5.12. In addition, there are various signal programs, such as four-stage control with protected turning phase, or the one which separates pedestrian from turning vehicle into different stages.

Signal control typically uses three signal control parameters: cycle length, green time split, and offset. "Cycle length" refers to the time it takes to complete one

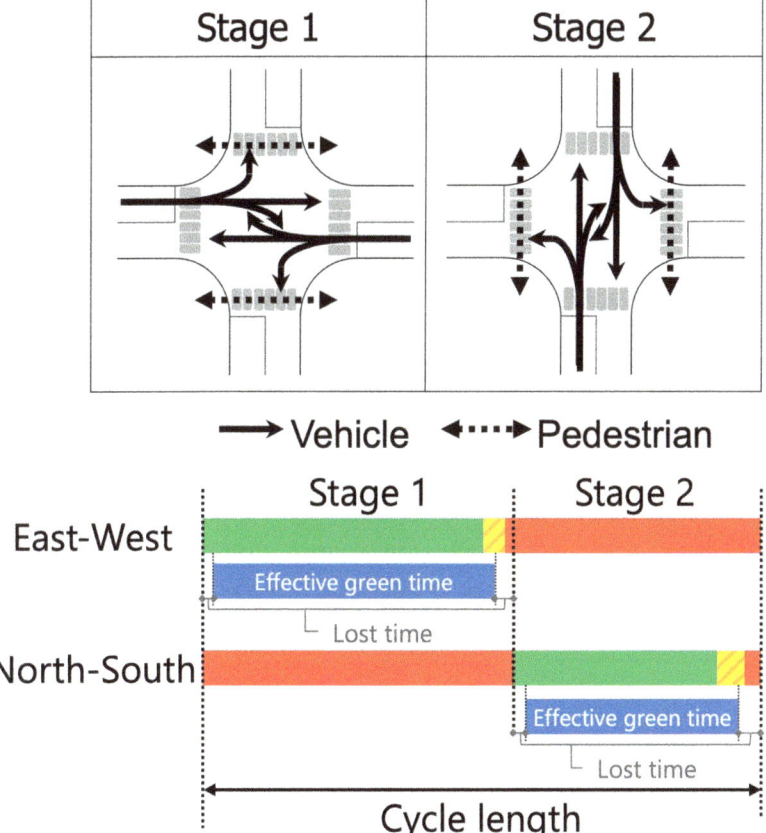

Fig. 5.12 Traffic signal control terms with an example

cycle, as shown in Fig. 5.12 (bottom). If there is a sufficient queue at the onset of the green time in each stage, generally, we can measure the maximum traffic flow rate, which is called "saturation flow rate" [veh/h/lane]. "Effective green time" is defined as the number of vehicles that passed the stop line during one green time [veh/lane] divided by the saturation flow rate, then "green time split" is defined as proportion of effective green time to cycle length. Besides, in "coordinated control" that coordinates adjacent intersections to each other, "(relative) offset" is defined as the difference in the onset of green times in the same direction between adjacent intersections.

At signalized intersections, delay occurs according to the signal control parameters even when traffic demand is zero. Also, traffic capacity of a signalized intersection is determined by its cycle length and green time split, and which tend to restrict traffic capacity of highways and streets as bottlenecks, particularly at critical intersections with high traffic demand. Therefore, an appropriate combination of geometric design (appropriate number of lanes and lane configuration inducing exclusive/shared lanes, introduction of splitter islands, etc.) and signal control (appropriate signal program and control parameters) is required. In recent years, research on new types of intersections called "Alternative Intersections" [13, 14] is also progressing, which can reduce traffic conflicts between turning and through vehicles by intersection geometry and signal control, thus improving efficiency and safety.

5.4 Measures for Enhancing Traffic Safety and Efficiency

Improvement of traffic safety would be achieved by a multifaceted approach, not only traffic engineering, which mainly focuses on infrastructure, but also vehicle technology as well as human factors. At least, in traffic engineering, as mentioned in Sect. 5.2.2, it is known that traffic congestion increases the risk of traffic crashes at the upstream end of queue. Also, safety is reduced when different types of road users, such as high-speed vehicles, access and egress vehicles, loading vehicles, bicycles, and pedestrians, have to share the same space and time due to unclear function/priority of the road. This section provides an overview of the measures that can alleviate traffic congestion and establish a functional hierarchical road network because these will also lead to improve safety. For the measures that directly address traffic crash reductions at specific locations such as accident-prone spots (black spots), see Chap. 6. For the measures using vehicle technologies such as ACC (Adaptive Cruise Control), see Chap. 7.

5.4.1 Road Traffic Congestion Measures

To eliminate or alleviate traffic congestion, it is necessary to either reduce traffic demand or increase bottleneck capacity. Traditionally, the former measures are called "Transportation Demand Management (TDM)," including measures that encourages road users to change their behavior such as modal shift, departure time, and route change, and efficient use of carpooling among automobile users. The latter mainly involved road constructions such as completing the network and increasing the number of lanes. In recent years, thanks to the enhanced traffic monitoring technology in a more real-time manner, both measures are being introduced more dynamically, as "Active Transportation and Demand Management (ATDM)" [15], mainly in Europe and the United States.

Typical measures of TDM are traffic information provision and road charging, which encourage changes in departure times or routes of automobiles. In addition, mainly in North America, there are measures to promote carpool by introducing high-occupancy Vehicle (HOV) lanes, which are dedicated for vehicles with a certain number of passengers (usually 2–3 persons or more), and high-occupancy toll (HOT) lanes, which charge tolls to vehicles that do not have the required number of passengers. In ATDM, a more real-time traffic information is provided, which may include predictions for the near future; charging prices change dynamically in response to changing traffic conditions, and the HOV/HOT lane qualifications are also dynamically determined. Not only HOV/HOT lanes but also lanes where operational strategies are dynamically changing are all called "managed lanes" (Fig. 5.13).

On the other hand, ATDM measures related to regulations and operations applied to specific bottleneck sections are categorized as Active Traffic Management (ATM), including ramp metering, variable speed limits, hard shoulder running, dynamic junction and merging control, and lane reversal. Ramp metering is a strategy to control inflow by traffic signals installed on expressway onramps to prevent congestion on the mainline and associated traffic crashes such as rear-end collisions.

Fig. 5.13 HOV lane

Variable speed limits (Fig. 5.14a) is a strategy to adjust the speed limit of each lane according to traffic conditions, road surface, weather conditions, and so forth on expressways. By applying the variable speed limits which lowering the speed limits from the upstream to the downstream where traffic volume is getting closer to the capacity, vehicles' speed can be harmonized, enabling to mitigate or prevent congestion and reduce the risk of rear-end collisions. Hard shoulder running (Fig. 5.14b) enables dynamic use of hard shoulders as an extra driving lane to increase traffic capacity when traffic congestion is severe. Dynamic junction control dynamically allocates lanes according to relative traffic demands of the mainline and ramps in interchange areas. Dynamic merge control (Fig. 5.14c) [17] provides drivers with advisory messages/signs of the merging and lane changing positions according to the traffic volume and merging ratio. Both junction and merge controls are expected to maximize traffic capacity in response to variable traffic patterns throughout the day and also prevent crashes caused by complex conflicts. Lane reversal is the strategy to operate a part of multiple lanes reversibly in response to traffic demand. It can efficiently allocate the capacity of congested roads where the directional ratio of traffic demand varies greatly by time-of-day. Examples can be seen on surface streets (Fig. 5.14d), expressway toll booths, etc. In Japan, light guidance systems (Fig. 5.14e) [16, 18] have been introduced at some sags and uphill sections, major bottlenecks on motorways, which can also be considered a type of ATM. The system consists of continuous installation of LED lights on the roadside and lighting

Fig. 5.14 Examples of ATM measures (**a**) Variable speed limits (**b**) Hard shoulder running (Left: closed hard shoulder, Right: opened hard shoulder) (**c**) Dynamic merging control [17] (**d**) Lane reversal (**e**) Light guidance systems [18]

5 Traffic Engineering

Fig 5.14 (continued)

Fig 5.14 (continued)

them up along with the driving direction while changing their blinking speed according to traffic conditions, expecting to alleviate congestion [16].

ATM enables the maximum utilization of traffic capacity realized by dynamic operation of an existing road infrastructure in response to traffic demand, which varies spatially and temporally. However, it is difficult to alleviate or eliminate traffic congestion only with such a dynamic adjustment in case when the amount of road network is not yet sufficient, for example, in developing countries, or in case when traffic demand far exceeds traffic capacity. In such cases, road construction such as enhancement of the road network connectivity and/or increase of the number of lanes is still necessary while maximizing these effects by introducing ATDM. It may be considered to increase the number of lanes by narrowing the lane width without changing the whole roadway width under spatial constraints.

5.4.2 Safety and Efficiency Measures According to Functional Hierarchy

To realize a functional hierarchical road network, it is important to maintain appropriate QoS according to the functions of roads. That means, the roads where through-movement function has higher priority require efficiency measures to maintain the speed of automobiles at a certain level, whereas the roads where the land-access and staying functions or non-motorized users such as pedestrians are highly prioritized require measures for traffic calming to deteriorate automobile speed.

Efficiency measures include not only the ones to eliminate or alleviate congestion mentioned in Sect. 5.4.1, but also the ones to improve QoS by reducing delay and providing the opportunities of overtaking. For that, at intersections, as mentioned in Sects. 5.3.3 and 5.3.4, it is necessary to select appropriate type of control such as grade separation and access management, and to well design road geometry and control details. As one measure of ATM, actuated control that dynamically adjusts green time based on the arrival of vehicles detected by traffic detectors (refer to Sect. 5.5.1) is also effective at signalized intersections. If the priority of the road is on public transport, introduction of Transit Signal Priority (TSP, sometimes called as "Public Transport Priority Systems (PTPS)") that adjust traffic signals according to the arrival of transit vehicles such as buses and associated road geometry should be actively discussed. On the other hand, the roads in mountainous areas and two-lane highways often require an extra lane for providing overtaking opportunities to maintain the certain QoS. The "2 + 1 road" is an example for such a measure, where the center lane of the three lanes is used alternately as an overtaking lane.

As traffic calming measures, in addition to applying the "Zone 30," which adopts the speed limit of 30 km/h or lower in the whole target area, physical devices can be installed to reduce vehicles' speed. These include speed humps (a raised section of road pavement), chokers (extended curbs that narrow the road space for vehicles), chicanes (deliberate S-shape curves in the road) [19], etc. Besides, "shared space" has been increased particularly in Europe [20], which intentionally does not define a boundary between sidewalk and roadway to make road users to pay more attention to others and create a safe space for pedestrians.

In any case, it is very important to implement a combination of traffic efficiency and calming measures systematically in an entire network or area, rather than to apply those individually, so as to meet the needs of all road users by promoting appropriate use of roads at different hierarchical levels according to their needs. Moreover, it is preferable that each road is designed and operated to make its function/priority naturally understood by road users (self-explaining road) through the road alignment, geometry, intersection allocation, etc., which encourages users to take safer actions spontaneously.

5.5 Traffic Observation and Prediction Technology

One of the major challenges in traffic engineering is the difficulty of conducting experiments in the laboratory. Therefore, in order to analyze traffic flow and develop road planning, design, and operation methodologies, it is essential to carefully observe and survey what happened on roads. In order to implement appropriate management in response to changing traffic conditions, it is necessary to predict traffic flow that is currently being realized/will be realized in the future. This section introduces technologies for traffic observation and prediction.

5.5.1 Traffic Observation

Traffic observation has been traditionally conducted manually by surveyors. However, thanks to the advancement of digitalization, it has been rapidly automated year by year. Traffic detection enables to detect the presence of vehicle using sensors mounted or embedded on roads and measure traffic volume and speed. These information are utilized to provide congestion information of expressways and to control actuated traffic signals. In recent years, detectors that use video cameras and image processing technology are increasing. Traffic detectors using image processing have advantages such as being able to observe multiple lanes at once and to observe traffic with no lane discipline, such as pedestrian traffic and heterogeneous traffic in developing countries. At the same time, they still have disadvantages in measurement accuracy, particularly in cases with high traffic volume, due to occlusion, where a large vehicle hides small vehicle(s) behind. For an observation of dedicated pedestrian spaces, technologies that cannot identify individuals, such as laser sensors, are preferred instead of video cameras.

Although these detectors and sensors have great advantages for traffic observations, these can be installed only in limited places, so there are problems that traffic conditions at all bottlenecks cannot be covered, and traffic conditions on roads without these detectors or sensors, typically on minor roads or streets, are unknown. Furthermore, to reduce traffic congestion and crashes, it is necessary to identify the locations and conditions where speed reduction as well as sudden acceleration and deceleration tend to occur, but such a spatial-temporal fluctuation of traffic flow can hardly be analyzed by the fixed detectors. In contrast, in recent years, advanced information and communication technology has established "probe" technology that can measure and monitor the traffic conditions at any given moment experienced by a specific vehicle or person with car navigation systems installed in that vehicle or a smartphone carried by that person. For example, in Japan, the collection of probe data using vehicle-to-infrastructure communication between vehicle onboard units and roadside antennas, called "ETC2.0," [21] has been deployed nationwide and is used to identify bottlenecks and locations with frequent sudden decelerations. However, it should be noted that probe data is essentially a sample, which cannot provide information about the quantity of the whole population.

5.5.2 Traffic Prediction Technology

Recently, technology has been developed [22, 23] to estimate the overall traffic conditions by fusing detector data, which provide information about the whole population at limited fixed locations, and using a probe data, which are a sample but can provide more detailed traffic conditions (example: Fig. 5.15) [24], based on the traffic flow theory explained in Sect. 5.1. Furthermore, by accumulating a large amount of detector and probe data and utilizing the latest data mining methodology, it is expected to drastically enhance our understanding of traffic conditions in the overall network. Such understanding makes it possible to estimate real-time traffic conditions accurately (Nowcast) [25, 26] through traffic simulations that are based on traffic theory and reflect historical and the latest detector and probe data. In the future, the application of various traffic management measures such as traffic information provision, route guidance, and any ATM measure like ramp metering can be

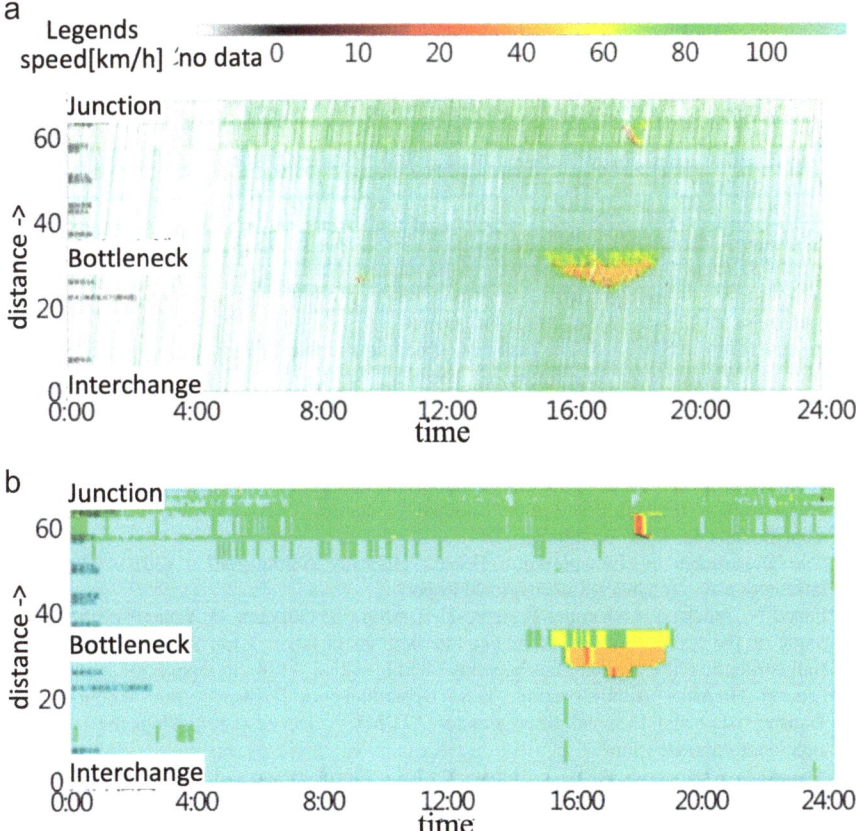

Fig. 5.15 Speed contour diagrams by probe data and vechicle detectors (**a**) speed contour derived by ETC2.0 probe data (**b**) speed contour derived by vehicle detectors [24]

tested in the digital twin that predict future traffic conditions. In such a digital twin, traffic crashes would be quickly detected based on information obtained from driving vehicles' sensors and swiftly implement traffic management measures such as deployment of ambulances, temporary lane closure, and detour guidance. This could potentially minimize the impact of crashes and second effects caused by the crashes. Furthermore, digital twin is expected to be greatly useful in planning and making actions for unpredictable incidents such as extremely bad weather, large-scale event, and others.

References

1. Koshi M (1985) Traffic flow phenomena in expressway tunnels. IATSS Res 9:50–56
2. Koshi M (1986) Capacity of motorway bottlenecks. J Jpn Soc Civil Eng 371/IV-5:1–7. (in Japanese)
3. Lighthill MJ, Whitham GB (1955) On kinematic waves: 2. A theory of traffic flow on long crowded roads. Proc R Soc Lond A 229:317–345
4. Koshi M, Akahane H (1988) Study on traffic congestion. Road Traffic Econ 45:64–69. (in Japanese)
5. Koshi M, Kuwahara M, Akahane H (1992) Capacity of sags and tunnels on Japanese Motorways. ITE J 62(5):17–22
6. Geroliminis N, Daganzo CF (2008) Existence of urban-scale macroscopic fundamental diagrams: some experimental findings. Transp Res B 42:759–770. https://doi.org/10.1016/j.trb.2008.02.002
7. World Health Organization. Global status report on road safety 2018. https://www.who.int/publications/i/item/9789241565684
8. Hyodo S, Yoshii T (2016) An analysis of the impact of hourly traffic volume on traffic accident risk on census highways. J Jpn Soc Civil Eng Ser D3 (Infrastructure Planning and Management) 72(5):I_1283-I_1291. (in Japanese)
9. Roger PR, Elena SP (2014) The highway capacity manual: a conceptual and research history volume 1: uninterrupted flow. Springer, Cham
10. Oguchi T (2008) Redesign of transport systems on highways, streets and avenues. IATSS Res 32(1):6–13
11. Japan Society of Traffic Engineers. Guidelines for Functionally Hierarchical Road Network Planning (Draft). https://www.jste.or.jp/Activity/h27-29.pdf, 2018. (in Japanese)
12. Transportation Research Board of the National Academies (TRB) (2014) Access management manual, 2nd edn. TRB, Washington, DC
13. U.S. Department of Transportation Federal Highway Administration (2010) Alternative Intersections/Interchanges: Informational Report
14. Inukai N, Tanaka S, Nakamura F, Ariyoshi R, Miura S, Oneyama H, Yanagihara M (2018) Study on the applicability of alternative intersections in Japan. J Jpn Soc Civil Eng Ser D3 (Infrastructure Planning and Management) 74(5):I_1327–I_1338. (in Japanese)
15. Federal Highway Administration, U.S. Department of Transportation. About Active Transportation and Demand Management (ATDM) – Overview, https://ops.fhwa.dot.gov/atdm/about/overview.htm
16. Kamata Y, Watanabe S, Anzai J, Shibata K (2013) Development and operation of spontaneous light-emitting pacemaker for traffic congestion measures. Proc Conf Jpn Soc Traffic Eng 33:181–184. (in Japanese)
17. Texas A&M Transportation Institute. Dynamic Merge Control, Transportation Policy Research. https://policy.tti.tamu.edu/strategy/dynamic-merge-control/

18. Metropolitan Expressway Co., Ltd. Sightline guidance and speed guidance system (escort light). Technology of Metropolitan Expressway. https://www.shutoko.jp/ss/tech-shutoko/use/escortlight.html. (in Japanese)
19. Japan Society of Traffic Engineers (2017) Revised Manual for Measures on Local and Residential Roads. (in Japanese)
20. Kojima A (2022) Efforts to apply shared space overseas and in Japan. IATSS Rev 46(3):203–210. (in Japanese)
21. ITS Technology Enhancement Association (ITS-TEA) About ETC2.0. https://www.its-tea.or.jp/english/its_etc/service_etc2.html
22. Mehran B, Kuwahara M (2011) Fusion of probe, passing time and signal timing data to estimate vehicle trajectories on urban arterials. Traffic Eng 46(1):77–89
23. Mehran B, Kuwahara M, Naznin F (2012) Implementing kinematic wave theory to reconstruct vehicle trajectories for fixed and probe sensor data. Transp Res C 22:144–163
24. Goto A, Matsuda N, Yamada K, Horiguchi R, Yoshida H, Sakaki S (2017) Case study of the traffic monitoring method for ring road network utilizing ETC2.0 probe data. Proc Conf Jpn Soc Traffic Eng 37:409–416. (in Japanese)
25. Hanabusa H, Kobayashi M, Koide K, Horiguchi R, Oguchi T (2013) Development of the Nowcast Traffic Simulation System for Road Traffic in Urban Area. In: Proceedings of 20th World Congress on ITS
26. Oguchi T, Chikaraishi M, Iijima M, Oka H, Horiguchi R, Tanabe J, Mohri Y (2018) Traffic management measures on Tokyo metropolitan urban expressway rings. J Jpn Soc Civil Eng Ser D3 (Infrastructure Planning and Management) 74(5):I_1255–I_1263. (in Japanese)

Open Access   This chapter is licensed under the terms of the Creative Commons Attribution-NonCommercial-NoDerivatives 4.0 International License (http://creativecommons.org/licenses/by-nc-nd/4.0/), which permits any noncommercial use, sharing, distribution and reproduction in any medium or format, as long as you give appropriate credit to the original author(s) and the source, provide a link to the Creative Commons license and indicate if you modified the licensed material. You do not have permission under this license to share adapted material derived from this chapter or parts of it.

The images or other third party material in this chapter are included in the chapter's Creative Commons license, unless indicated otherwise in a credit line to the material. If material is not included in the chapter's Creative Commons license and your intended use is not permitted by statutory regulation or exceeds the permitted use, you will need to obtain permission directly from the copyright holder.

Chapter 6
Technology for Improving Safety: Infrastructure Part

Koji Suzuki and Hideki Nakamura

6.1 Improving Safety at At-Grade Intersections [1]

At at-grade intersections, various users such as cars, bicycles, and pedestrians from multiple directions use the same space, causing their paths to intersect and creating a road facility where safety issues are likely to occur. Moreover, the quality of design and control of at-grade intersections not only affect safety but also significantly impact the efficiency of traffic. Therefore, when considering design and control, it is required to highly balance both safety and efficiency.

In this chapter, the principles of intersection structure and roundabouts are discussed, as methods of handling at-grade intersections to achieve both of these goals.

6.2 Principles of At-Grade Intersections

6.2.1 Number of Legs at At-Grade Intersections

As a rule, at-grade intersections should not have more than five legs. Although it depends on the traffic control method, the number of conflict points where traffic flows intersect, merge, and diverge at at-grade intersections sharply increases as the number of legs does, and the attention and judgment required of drivers also increase, raising the level of danger. At such signalized intersections, the number of

---

K. Suzuki (✉)
Department of Architecture, Design, Civil Engineering and Industrial Management Engineering, Nagoya Institute of Technology, Nagoya, Japan
e-mail: suzuki.koji@nitech.ac.jp

H. Nakamura
Department of Environmental Engineering and Architecture, Nagoya University, Nagoya, Japan

© The Author(s) 2026
Pioneering the Future for Traffic and Safety Sciences,
https://doi.org/10.1007/978-981-96-0676-4_6

signal displays increases to handle complex traffic flows, and the total loss time during signal switching increases, reducing capacity and potentially causing significant delays.

At roundabouts, even if there are many legs, the number of conflict points does not increase sharply like ordinary intersections, and if traffic demand is low, it may reduce delays compared to other traffic control methods. However, there are many points to note, so careful consideration is required when applying them.

6.2.2 Shape of At-Grade Intersections

6.2.2.1 Crossing Angle Between Legs

Generally, at intersections, the crossing angles between legs must be planned to be at right angles or close to it. If the crossing angle between legs is less than 90°, it is called an acute angle, and if it is larger, it is called an obtuse angle (Fig. 6.1). For example, as shown in Fig. 6.1b, the direction of a vehicle turning right at an acute angle will have to turn more than 90° (the turning angle increases). This is also the case for left-turning vehicles when the inflow and outflow are reversed. At right-angle or near-right-angle intersections, the length of the intersecting roadway is short, and the area of the intersection is also small. If the crossing angle between legs becomes acute or obtuse, visibility can become extremely poor depending on the direction, and the distance between stop lines will increase. All of these factors affect traffic safety and capacity.

6.2.2.2 Irregularly Shaped Intersections

Irregularly shaped intersections with special shapes, such as "staggerd" intersections where the center lines of opposing inflow sections do not intersect at one point, or "bent leg" intersections where the center lines of opposing legs are bent, have safety and efficiency issues. Therefore, it is desirable to avoid them. Such intersections have large crossing areas, and the trajectories of each traffic flow become

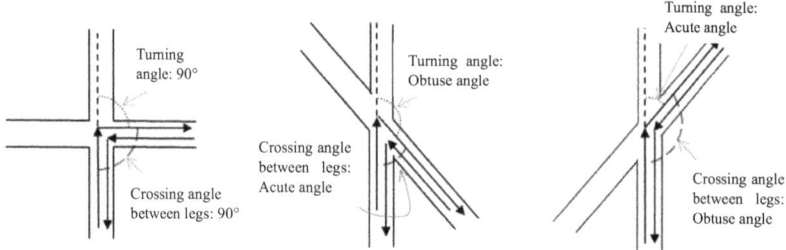

Fig. 6.1 Crossing angle between legs

complex. Furthermore, these trajectories intersect with those of pedestrians, reducing both safety and capacity.

6.2.3 Compact Intersections

Reducing the size of intersections (making intersections compact) is important from both safety and efficiency perspectives, as it shortens the crossing distance and time for pedestrians, improves safety, and, if signal-controlled, reduces lost time and expands intersection capacity. Furthermore, from a safety perspective, there are also the following benefits to making intersections compact.

- The total red time can be shortened, which can prevent the induction of the red-light running at times such as signal switching.
- The range where vehicles, bicycles, and pedestrians conflict is limited, the variation in vehicle trajectories is reduced, and the range where drivers should concentrate their attention can be clarified.
- The exiting speed of right- or left-turning vehicles, which is the speed when passing through a crosswalk, can be controlled.

The size of an intersection can be easily understood by measuring the distance between opposing stop lines (Fig. 6.2). The distance between stop lines is determined by factors such as the curve radius of the corner, the crosswalk width, and the separation between the crosswalk and the stop line. Therefore, when designing these elements, care must be taken to avoid creating an excessively large intersection.

On the other hand, if the separation to the crosswalk at the left-turn exit is small particularly in the case of shared left-turn lanes, left-turning vehicles waiting for

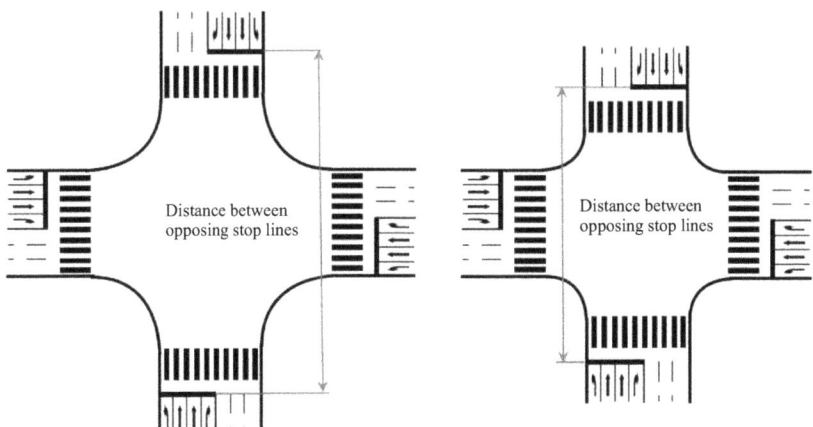

Fig. 6.2 Concept of compact intersection

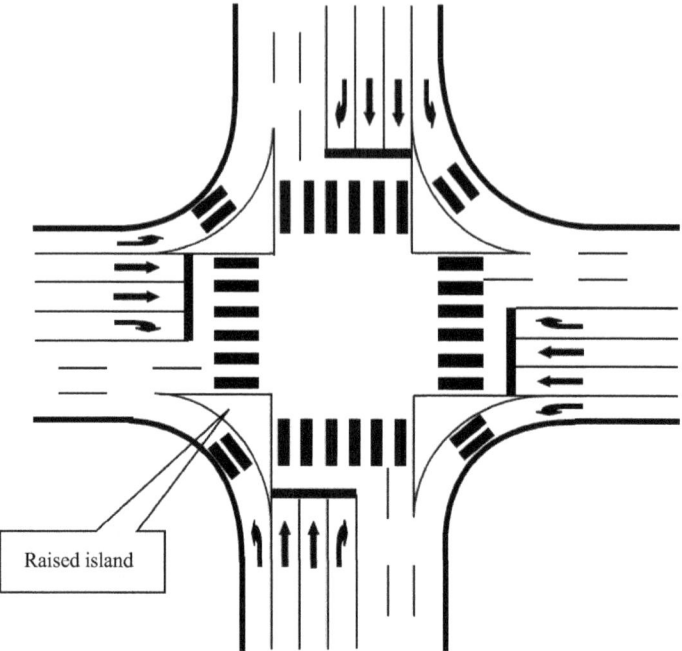

Fig. 6.3 Concept of channelization at intersection

pedestrians to cross may obstruct the passage of following through vehicles. This can result in reducing the lane capacity and needs to be taken into account.

Furthermore, as shown in Fig. 6.3, by providing a raised island within the intersection and channelizing the left-turn traffic flow (channelization), the intersection can be significantly compacted. However, in this case, there is a concern about the conflict between left-turning vehicles and pedestrians crossing on the left-turn channelized path. Therefore, when determining the shape of the channelized left-turning lane and the position of the crosswalk, it is necessary to ensure sufficient visibility and to implement measures to draw attention of left-turning drivers.

6.3 Roundabouts [2]

6.3.1 Definition of a Roundabout

A roundabout is a circular at-grade intersection that allows vehicles to travel clockwise (right turn) on the circulatory roadway primarily by arranging certain structural elements, as shown in Fig. 6.4, and ensures unobstructed circulatory traffic that is not hindered by entering vehicles. In roundabouts, the vehicles on the circulatory

6 Technology for Improving Safety: Infrastructure Part

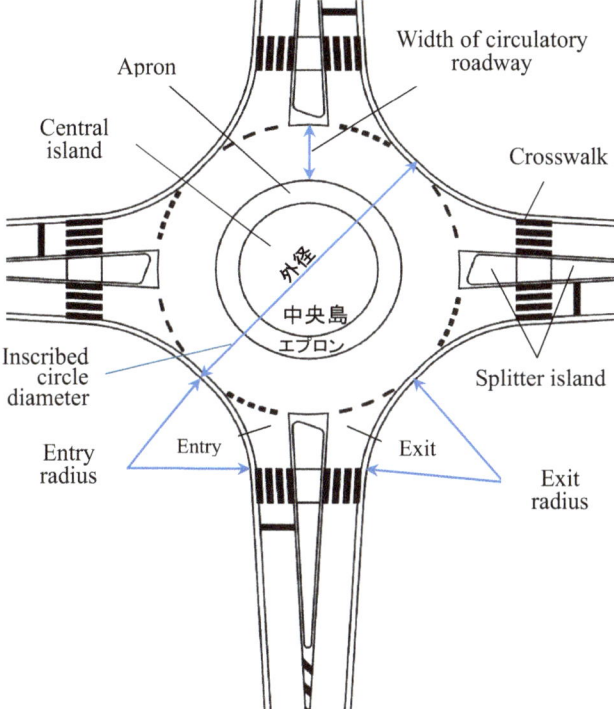

Fig. 6.4 Structural elements of roundabout

roadway have priority over the entering vehicles, and the traffic flow on the circulatory roadway must not be interrupted by traffic signals or stop signs.

6.3.2 Characteristics of a Roundabout

6.3.2.1 Improved Safety at At-Grade Intersections

Roundabouts can reduce the number of conflict points between vehicles within the intersection. Figure 6.5 shows the number of conflict points at a typical four-leg intersection. While a regular unsignalized intersection has twenty conflict points, a roundabout significantly reduces this to four. Also, in roundabouts, all entries are non-priority controlled, and due to the physical structure of having a central island, the inflow speed is suppressed, and collisions between vehicles such as head-on or right-turn versus straight-on, which cause significant damage, do not occur. For these reasons, even if a traffic crash occurs, it is characterized by a form of crash with less damage.

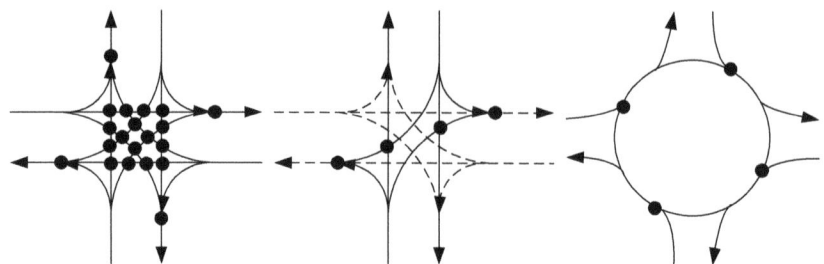

Fig. 6.5 Number of conflicts points at a typical four-leg intersection

6.3.2.2 Improving Intersection Efficiency by Reducing Delays

At signalized intersections, even if there are no vehicles in the crossing direction during the red light, one must continue to wait for the light to turn green. On the other hand, at roundabouts, if there are no vehicles traveling on the circulatory roadway, it is possible to enter the intersection at any time. This can significantly reduce delays, especially during off-peak hours, and consequently greatly reduce the overall travel time within the relevant section.

6.3.2.3 Improving the Processing Efficiency (Capacity) of Irregularly Shaped Intersections

Intersections with five or more legs have many conflict points compared to regular four-leg intersections. However, the introduction of roundabouts can significantly reduce the number of these conflict points. Also, when signal control is implemented at irregularly shaped intersections such as multi-branch intersections and/or staggered intersections, complex signal settings are generally required, which inevitably reduces the green time ratio per entry and increases delays. At roundabouts, regardless of the number of legs at the intersection, it is possible to significantly reduce delays particularly when traffic demand is low.

6.3.2.4 No Need for Implementing Turning Lanes

At roundabouts, as all directions of traffic (through, left-turn, right-turn) can enter from the same entry lane, there is no need to implement right-turn and left-turn lanes at the entry.

6.3.2.5 Lower Lifecycle Costs and Environmental Impact

Roundabouts can operate intersections without using electricity, other than for road lighting.

6.3.2.6 Roundabouts Resilient to Disasters

In the case of roundabouts, even if the intersection is damaged by a natural disaster such as a tsunami, as long as the road structure itself is not significantly damaged, it can function "autonomously" without relying on electricity or manpower, and it is expected to operate safely and efficiently, almost as usual.

6.3.3 Points to Note

Comparing to properly designed regular at-grade intersections, the entry capacity of roundabouts is generally lower than that of signalized intersections. Therefore, it is not possible to apply roundabouts to intersections with high traffic demands or to introduce roundabouts for the purpose of traffic congestion measures. However, in cases where high capacity cannot necessarily be secured even with signal control due to the intersection being an irregularly shape, such as a multi-leg or staggered intersection, there may be cases where capacity can be increased by converting to a roundabout.

Also, it is necessary to pay close attention to ensuring the safety of pedestrians and bicycles. As a safety measure for crossing pedestrians, it is desirable to adopt a two-stage crossing method at crosswalks.

6.3.4 Types of Roundabouts

Roundabouts are generally classified into compact roundabouts, multi-lane roundabouts, mini roundabouts, etc., based on the number of lanes in the entry/exit sections and the circulatory roadway, and whether or not it is possible to ride up to the central island. Recently, the number of turbo roundabouts, which is a form developed in the Netherlands, has been increasing in Europe and elsewhere.

6.3.5 Significance of Introducing Roundabouts

The main reasons for introducing roundabouts in Japan are (a) improving safety at unsignalized/signalized intersections where serious accidents occur due to head-on collisions or right-turn versus oncoming through vehicles crashes, (b) calming traffic by controlling vehicle speeds at at-grade intersections in residential areas, and (c) reducing the waste of control delays caused by traffic signals at intersections with low traffic demand despite being signalized.

On the other hand, when considering improvements to existing at-grade intersections for safety and smooth traffic flow, it is generally believed that there are significant constraints on space conditions.

Therefore, in Japan, so far, from the perspectives of safety improvement effects due to vehicle speed control, space constraints, and right-turn methods, "compact roundabouts" with a structure where the circulatory roadway has one lane, and both the entry and exit connecting to the circulatory roadway have one lane, and vehicles cannot ride up to the central island, have been introduced.

6.4 Improving Safety of Basic Segments

Accidents involving pedestrians, especially the elderly, crossing at parts of the road other than intersections and basic segments, remain a problem to be solved, and consideration of measures is required. In this section, two-stage crossing facilities as safety measures at crossing points and physical devices on residential roads that contribute to accident reduction by speed control are discussed.

6.4.1 Two-Stage Crossing Facilities [3]

There are survey results indicating that many accidents involving elderly people crossing roads occur with vehicles approaching from the left. As a mechanism to allow pedestrians to cross safely and easily, there is a focus on the "two-stage crossing" method, which involves dividing the crosswalk by providing a space in the center of the road by installing traffic islands or median strips, allowing the road to be crossed in two stages. Overseas, the two-stage crossing method is widely used in both urban and rural areas, regardless of whether it's at basic segment or intersection, as an alternative to traffic lights or footbridges. In our country, the spread of this crossing method is expected to realize a safe and comfortable road space. In this section, an overview of the features of two-stage crossing facilities and the points to consider in their application are discussed.

6.4.1.1 Two-Stage Crossing and the Features of Two-Stage Crossing Facilities

Two-stage crossing is a method where a traffic island is installed in the middle of the roadway for pedestrians to wait and take refuge, and the crossing is divided into two stages before and after this island. The target of the crossing pedestrians here includes walking the bicycles, wheelchair users, etc. Two-stage crossing has both unsignalized and signalized methods, each classified into basic segments and

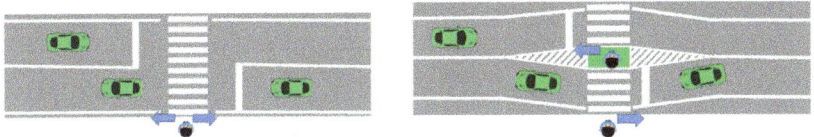

Fig. 6.6 General crosswalk (left) and two-stage crossing (right)

at-grade intersections, but this chapter will explain examples of unsignalized methods focusing on infrastructure.

In Japan, crosswalks on basic segments typically are not divided by placing a traffic island in the middle of the road, as shown on the left in Fig. 6.6. In contrast, as shown on the right in Fig. 6.6, when a refuge space is installed in the middle of the road by a traffic island, and the crosswalk is divided into two by this, it is as if crossing a narrow one-way road twice in the case of two-stage crossing. That is, first, start crossing after confirming the safety of the right direction, then you can confirm the vehicles coming from the left on the refuge space in the center. This allows you to judge the safety of each direction individually. This not only makes it easier to judge and increases crossing opportunities but also shortens the crossing distance on the roadway required for one crossing. In addition, the presence of a traffic island makes it easier for drivers to pay attention to the crossing area, and if there are pedestrian on the traffic island, it is easier for drivers to recognize their presence.

The installation of this traffic island is expected to have the effect of reducing vehicle speed near the crossing area. In addition, compared to pedestrian push-button signals, etc., the effect of shortening the waiting time for crossing and the effect of clearly conveying such changes to road users by installing two-stage crossing facilities at locations where road hierarchy divisions change on the same route or where regional characteristics change along the road can be expected. Furthermore, it is advantageous in terms of reducing the burden of facility maintenance compared to installing new traffic lights for the safety of crossing pedestrians. Another difference from crossing facilities with traffic lights is the ability to ensure crossing capacity during power outages.

6.4.1.2 Basic Structure of Two-Stage Crossing Facilities

Two-stage crossing facilities are composed of a traffic island and a crossing section, as shown in Fig. 6.7.

The traffic island is a facility, installed in the middle of the crossing section to ensure the safety of pedestrians crossing. The crossing section is the part where pedestrians cross. A shift section may be provided in cases where it is necessary to shift the main line in the sections before and after the traffic island.

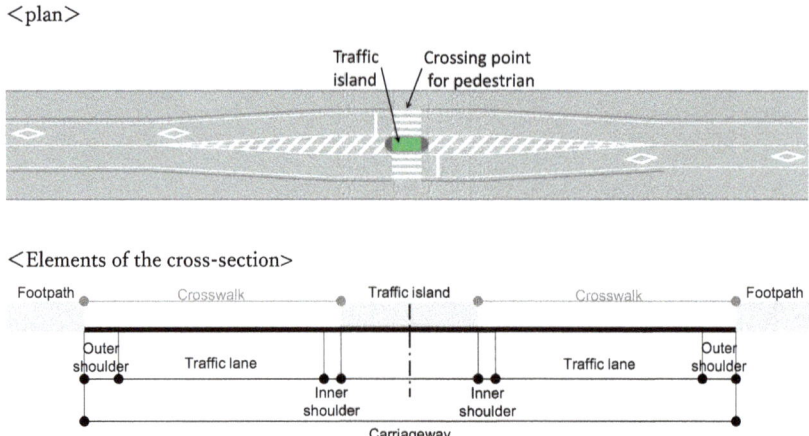

Fig. 6.7 Basic structure of two-stage crossing facilities

Considering domestic and international examples, the types of two-stage crossing facilities without traffic lights are classified by the combination of the operation and structural form of the crossing section. The operation is classified into whether to install a crosswalk or pedestrian crossing guidance line in the crossing section, or whether to not specify the crossing position. The structural form of the crossing section is classified into whether to make the crossing section straight or staggered.

Regarding the operation of the crossing section, in the type where a crosswalk is installed, vehicles are obliged to stop when there are pedestrians intending to cross. In the type where a pedestrian crossing guidance line is installed in the crossing section or the crossing position is not specified without marking, there is no regulation obliging vehicles to stop even when there are pedestrians intending to cross. There are no examples of the type that does not specify the crossing position in Japan, but this type is often applied in Western countries. It is applied in various situations, such as on arterial roads that prioritize efficiency, or when increasing and dispersing the number of places where crossing is possible on streets, etc., because pedestrians do not have priority.

Also, as a structural form of the crossing section, by making it a staggered structure, there are merits such as preventing the risk of pedestrians crossing all at once without checking vehicles at the traffic island and making it easier to check vehicles because they will face the direction from which vehicles approach when passing through the traffic island. On the other hand, pedestrians need a certain amount of space to move through the traffic island, and there is also a possibility of inducing jaywalking, such as when it becomes a detour for pedestrians or when they walk outside the crosswalk. In Western countries, it is more common for the crossing section to be straight, and the staggered type is often installed in special cases where vehicles need to be more careful.

6.4.1.3 Points to Note When Planning Two-Stage Crossing Facilities and Desirable Installation Locations

When considering a two-stage crossing facility as one of the options at the planning stage of a crossing facility, it is necessary to fully understand the characteristics of the two-stage crossing facility and evaluate its applicability.

Generally, situations where a crossing facility is required include cases where an increase in crossing demand is expected due to roadside development or residential location, cases where a road divides a region and there is a certain amount of crossing demand despite the absence of a nearby crossing facility, and cases where there is a risk of accidents involving pedestrians crossing outside of crosswalks or intersections. On the other hand, the road sections where crossing facilities are installed each have different functions, such as roads that prioritize "mobility function," sub-arterial roads and residential roads that prioritize "roadside access function (access function/stay function)." In addition, sub-arterial roads and residential roads can be subdivided into roads that form the backbone within the district to access the arterial roads and local roads that provide service functions to dwellings, etc., and have a hierarchical structure. Therefore, in planning a crossing facility, it is necessary to consider the functions of each road and their hierarchy. The places where the introduction of a two-stage crossing facility is desirable are classified as follows (see points a–d):

(a) Locations where the accident risk for crossing the road is high

The implementation of countermeasures at locations with frequent pedestrian–bicycle crossing accidents and clearly identifiable risks, as well as at locations where pedestrians frequently cross roadways outside designated crossing facilities, is effective.

(b) Locations where proper use of roads is required according to the roadside environment

At locations with high crossing demand, particularly during peak hours—such as railway stations, bus stops, commercial facilities, and residential areas—accident risk increases as pedestrians and vehicles, reluctant to accept longer crossing waiting times, force their way through traffic. Also, the crossing waiting time for vehicles increases and the efficiency of driving decreases as pedestrians intermittently cross the road. The introduction of a two-stage crossing facility at such places is expected to have effects in terms of both safety and efficiency. An example of introduction is the case of the unsignalized crosswalk on Prefectural Road 400, which connects to Kasukabe Station on the Tobu Isezaki Line in Kasukabe City, Saitama Prefecture (Fig. 6.8).

(c) Locations where it is necessary to solve road structural issues

In places where zebra markings are set up in the middle of the roadway, or in temporary operational sections and other places where there is room in the width, and in places where right and left turn processing and crossing processing are combined, the roadway width is wide, so the distance for pedestrians to cross the

Fig. 6.8 Two-stage crossing facility in Kasukabe City, Saitama Prefecture

roadway becomes long. This requires a longer crossing time and a higher need than usual to carefully determine the time required for the arriving vehicles and the time interval between arriving vehicles. This situation is particularly dangerous for elderly pedestrians with low walking speeds, especially when vehicle speeds are high, raising concerns about the occurrence of serious accidents. Introducing two-stage crossing facilities at locations with long crossing distances is effective, as such facilities shorten the crossing distance and exposure time within roadway spaces where conflicts with vehicles may occur and facilitate pedestrian decision-making.

(d) Places where it is desirable to introduce with contribution of urban development

The two-stage crossing facility not only ensures the safety of crossing pedestrians but also has functions such as clearly indicating that it is a point where the road hierarchy switches from arterial road with a car-priority to residential road with a pedestrian/bicycle-priority, and encouraging speed control of vehicles by changing the road alignment using a traffic island. Therefore, it is expected to utilize the two-stage crossing facility when considering urban space development, such as clarifying the road hierarchy and optimizing the use of roads by devising the road structure, rather than simply positioning the road as a route for vehicles and crossing pedestrians. As shown in the overseas example in Fig. 6.9, such installations can serve as symbolic gateway features when applied in contexts emphasizing pedestrian-priority urban design, such as entrances to tourist attractions and roadside stations, or major entry points to school zones and Zone 30 areas. Their application is recommended, taking roadway and regional characteristics into account.

6.4.1.4 Points to Note When Considering the Introduction of a Two-Stage Crossing Facility

In order to manifest the effects of the two-stage crossing facility and not to cause new issues or problems, the following points a–e should be noted when considering the introduction.

6 Technology for Improving Safety: Infrastructure Part 115

Fig. 6.9 Two-stage crossing facility in Hoofddorp, the Netherlands

(a) Confirmation of the possibility of installing a traffic island within the road width

It is necessary to confirm whether a traffic island can be installed within the current road width or by redistributing the road space.

(b) Confirm the impact on roadside entries and exits due to the introduction of a two-stage crossing facility

It is necessary to confirm whether the installation of traffic islands and other facilities as a two-stage crossing facility will not hinder the movement of vehicles entering and exiting narrow streets or roadside facilities.

(c) Confirmation of Pedestrian Characteristics

When considering a staggered crossing structure as a two-stage crossing facility, if the desired route of the crossing pedestrian differs at the application site, the pedestrian is forced to detour more than necessary, resulting in jaywalking and compromising safety during crossing. Therefore, it is necessary to understand the characteristics of pedestrian traffic volume and flow (OD), pay attention to the presence of vulnerable road users such as commute students, the elderly, and cyclists, and plan the structure of the crossing section.

(d) Impact on Vehicle Efficiency Reduction

When introducing two-stage crossing facilities, their impact on vehicular efficiency should be carefully considered, as the balance between pedestrian and vehicular traffic volumes may lead to reduced vehicle capacity. In two-stage crossing facilities equipped with crosswalks, vehicles are required to yield to crossing pedestrians; therefore, special caution is necessary on arterial roads with a primary function of traffic mobility.

(e) Consideration of Installation Locations in Relation to Adjacent Crosswalks

When planning the introduction of a two-stage crossing facility, it is important to view it as effectively functioning as a road network for the entire region. Therefore, it is necessary to consider the location of the introduction in relation to adjacent signalized intersections and crosswalks.

6.4.2 Physical Devices as Safety Measures on Residential Roads [4]

Some of the safety measures on residential roads are soft measures such as speed limit regulations of 30 km/h, parking prohibition, and prohibition of large vehicles, and hard measures such as installation of physical devices. The main objectives of physical devices related to infrastructure technology include traffic volume control, speed control, and securing pedestrian space.

Physical devices that contribute to traffic volume control include rising bollards and barricades installed at the entrance of blocks. In Europe, hard steel rising bollards are used, while in Japan, the use of soft rising bollards made of flexible materials has been seen in recent years [5].

For speed control, measures such as humps that provide convex parts on the road surface, smooth crosswalks, chicanes that make the planar shape of the road crank-shaped or S-shaped, and narrowing the width of the vehicle passage are mentioned.

Regarding the installation of humps on basic segments, continuous installation is necessary for speed reduction, and it is desirable that the installation interval be less than 40 m. However, caution is needed as problems such as excessive vehicle shaking can occur if the installation interval is too close. Also, there is a risk of complaints from citizens due to the generation of noise and vibration and the sound of re-acceleration when the vehicle speed is too high or there are many large vehicles. Consideration of these points is desirable when considering the installation interval. There are two types of hump shapes: trapezoidal, which has a flat top, and bow-shaped, which does not have a flat top. The bow-shaped hump is said to have a greater speed reduction effect than the trapezoidal hump, but it is prone to noise and vibration when the vehicle speed exceeds 30 km/h, so attention should be paid to this point. Also, on routes where buses pass, there is a risk that the bow-shaped hump may contact the bottom of the vehicle and the road surface, so it is desirable to use a trapezoidal hump. By installing a trapezoidal hump at the crosswalk, a smooth crosswalk is formed. This structure not only ensures the safety of pedestrians and wheelchair users by reducing vehicle speed, but also makes it easier for the elderly, strollers, wheelchair users, and others with mobility restrictions to cross the road. When installing, it is important to note that the sidewalk should be of the mount-up type as a basic rule. The height difference between the hump and the sidewalk edge should be 2 cm, the height of a flat part should be 10 cm as standard, and the sidewalk surface connected to the smooth crosswalk should be processed to meet the standards for barrier-free. In the single-lane section, the narrowness is expected to have an effect on speed reduction in the section due to continuous installation and combination with humps. The narrowness is formed by narrowing the outer line of the roadway, protruding the sidewalk, planting trees, and installing bollards. When forming a narrow part only with the outer line of the roadway, it is desirable to install rubber poles such as post cones and pole cones on the outer line of the roadway, and to make it clear and emphasize the narrow part, and to make it recognizable that the width of the roadway is changing. The narrow width of the narrow part of the roadway is set as standard at 3 m. In order to increase the traffic volume and speed reduction effect, it is possible to further narrow the width of the narrow part by limiting the passing vehicles, but in this case, the minimum width should not be less than the maximum vehicle width plus 0.5 m. In addition, the structure should consider the passage of emergency vehicles. In addition, it is required to secure more than 1 m as a pedestrian passage space in the narrow part and to consider the passage of bicycles. The shape of the narrowing can be selected to protrude on one or both sides depending on the roadside and traffic conditions. On roads with two-way traffic, a narrowing is particularly expected to reduce speed when a meeting with a vehicle coming from the opposite direction occurs. There is also a method of making it a one-sided narrowing so that it is clear which vehicle will yield. When making it a one-sided narrowing, consideration is given to the direction of the vehicle facing the pedestrian, and efforts are made to improve the sense of safety for pedestrians on the side without protrusion as a principle. In the case of a narrowing near the intersection of a two-way traffic road, a reduction in speed can be expected particularly when a meeting with a vehicle coming from the opposite direction occurs. However, since there is a possibility that vehicles may be held up within the intersection, a narrowing is not provided directly at the intersection in the case of two-way traffic. In the case of a one-way exit, it is basic to provide a narrowing on both sides at the intersection. When installed at the entrance intersection of a zone with a maximum speed of 30 km/h, it also contributes to improving the visibility of both drivers and pedestrians as the pedestrian crossing distance becomes shorter. Also, when installed at the zone boundary, an effect of emphasizing the zone boundary can be expected. To secure pedestrian space, measures such as pavement coloring to enhance pedestrian separation, the installation of rubber poles, and the two-stage crossing facilities described in the previous paragraph can be positioned as effective countermeasures.

6.5 Traffic Safety Measures Considering Both Arterial Roads and Residential Roads

Sections 6.1, 6.2, and 6.3 discussed infrastructure technologies that contribute to improving the performance of at-grade intersections, while Sect. 6.4 explained infrastructure technologies related to safety measures on basic road segments. However, in reality, there are many situations where measures that consider both are required. In this section, safety measures that consider both arterial roads and residential roads are discussed. Congestion on arterial roads can lead to shortcut traffic on residential roads, significantly reducing the safety of residential roads. To solve such problems, it is necessary to clearly distinguish between arterial roads that prioritize vehicle drivability and residential roads used by community people and to implement traffic safety measures that consider both. On arterial roads, drivability can be enhanced by reducing intersection size and revising signal control, as discussed above. In contrast, at interfaces between arterial and residential roads, it is desirable to discourage through traffic into residential areas by strategically implementing physical measures—such as roadway narrowings, continuous pedestrian crossings, two-stage crossing facilities, chicanes, and speed humps—to control vehicle speeds. To promote such approaches, comprehensive traffic safety measures involving coordinated planning and implementation by road administrators are being advanced in Japan [6].

References

1. General Incorporated Association Traffic Engineering Research Society, Planning and Design of At-grade Intersections – Basic Edition, 2018
2. General Incorporated Association Traffic Engineering Research Society, Roundabout Manual 2021, 2021
3. General Incorporated Association Traffic Engineering Research Society, Independent Research "Guide to Introducing Unsignalized Two-Stage Crossing Facilities (Draft)", 2021
4. General Incorporated Association Traffic Engineering Research Society, Revised Manual for Zone Measures on Living Streets, 2017.
5. Environmental Safety Division, Road Bureau, Ministry of Land, Infrastructure, Transport and Tourism "Rising Bollard Case Collection 2018"
6. Homepage of Nagoya National Highway Office, Ministry of Land, Infrastructure, Transport and Tourism, https://www.cbr.mlit.go.jp/meikoku/activity/safety/content02.html

Open Access This chapter is licensed under the terms of the Creative Commons Attribution-NonCommercial-NoDerivatives 4.0 International License (http://creativecommons.org/licenses/by-nc-nd/4.0/), which permits any noncommercial use, sharing, distribution and reproduction in any medium or format, as long as you give appropriate credit to the original author(s) and the source, provide a link to the Creative Commons license and indicate if you modified the licensed material. You do not have permission under this license to share adapted material derived from this chapter or parts of it.

The images or other third party material in this chapter are included in the chapter's Creative Commons license, unless indicated otherwise in a credit line to the material. If material is not included in the chapter's Creative Commons license and your intended use is not permitted by statutory regulation or exceeds the permitted use, you will need to obtain permission directly from the copyright holder.

Chapter 7
Technology for Safety Improvement: Automobile Edition

Taro Sekine

This chapter discusses the technical aspects of improving safety for automobiles. In Sect. 7.1, we look back at the trends and transitions in automobile safety technology development mainly in Japan. We introduce the experimental safety vehicle (ESV), advanced safety vehicle (ASV), intelligent transport systems (ITS), and current efforts toward the social implementation of autonomous driving.

In Sect. 7.2, we introduce the features of individual devices that have been put into practical use for active safety, and in Sect. 7.3, for passive safety.

In Sect. 7.4, we introduce accident prevention technology for elderly drivers, which has become a problem in recent years, and new safety improvement technology using connected technology.

7.1 Trends in the Development of Automobile Safety Technology

7.1.1 Dawn of Measures for Safety Issues: ESV Project

In the 1960s, traffic accidents were a problem not only in Japan but also in other countries due to the spread of motorization. To improve this situation, the experimental safety vehicle (ESV) development plan was proposed in 1970 by the NHTSA (National Highway Traffic Safety Administration) of the U.S. Department of Transportation (DOT) with the aim of producing experimental vehicles at the highest level of technology at the time (State of the-art Technology). The ESV project targeted the pursuit of safety and technological improvement for occupant protection and driver hazard avoidance, with the aim of reducing the number of deaths in

---

T. Sekine (✉)
Department of Mechanical Engineering, College of Science and Technology,
Nihon University, Tokyo, Japan
e-mail: sekine.tarou@nihon-u.ac.jp

vehicles. At that time in the United States, 60% of traffic fatalities were in cars, so it is presumed that the U.S. plan was primarily aimed at improving occupant protection in collisions [1].

In November 1970, a memorandum was exchanged between the U.S. government and the Japanese government and the then West German government, and Japan also participated in the ESV project. At this time, the U.S. was planning for a vehicle weight of about 1800 kg. On the other hand, in Japan, the specifications were examined by a special committee setup, and in 1971, two specifications were decided for a small car with a vehicle weight of about 1150 kg for a four-seater and about 900 kg for a two-seater, and development manufacturers were recruited. The Japanese specifications included targets such as ensuring survivable impact values and survival space for occupants even in collisions or rear-end collisions from 80 km/h, and also incorporated preventive safety technology leading to accident avoidance. Another notable feature is that vehicles were required to comply with both the Japanese Road Transport Vehicle Act and the U.S. Federal Motor Vehicle Safety Standards (FMVSS).

The following are some examples of safety technologies developed in the ESV project.

First, at 80 km/h as a technology to improve the impact absorption performance during full-frontal collisions, there is an improvement in the vehicle body structure. The frame for absorbing collision energy and pillars and side beams to secure the survival space for passengers can be mentioned [2]. At this time in the United States, the finite element method (FEM) application NASTRAN, which was transferred free of charge from the Apollo project by NASA, was introduced for frame structure calculations. It is noteworthy that the technology commonly used for body rigidity, calculations, etc. today was also used for automobiles at that time.

Next, the airbag system can be mentioned. The idea of collision detection technology and the airbag itself, which was called a safety cushion, existed before, but it was not practical. In the United States at the same time as the ESV project, there was a discussion on the legislation to mandate seat belt wearing, and the demand for passive restraint devices that can mitigate the impact of accidents even without seat belts increased, and the airbag system was considered in the ESV project. As a result, in 1974, General Motors (GM) became the first in the world to sell mass-produced cars equipped with an airbag system.

Also, in relation to accident-avoidance performance, a system was developed to improve vehicle stability during braking by using anti-skid brakes (current ABS) and vehicle lateral slip restriction function (current ESC).

The ESV project effectively ended with the development and performance evaluation of experimental vehicles by participating companies and the 5th International Conference held in 1974. The ESV achieved results by actually manufacturing experimental vehicles with the highest level of technology at the time but did not consider productivity and cost. Therefore, the energy and environmental issues that became problems during that time will be taken over by the RSV (Research Safety Vehicle) project, which assumes consumption trends and mass production [3]. It is

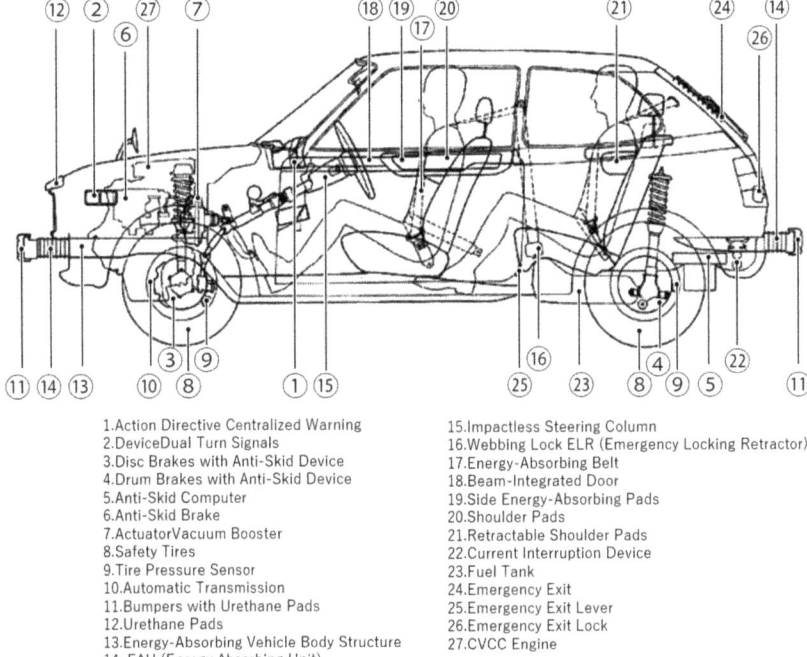

Fig. 7.1 Experimental safety vehicle (Honda ESV structural drawing) [4]

clear that the ESV project became a milestone in the development of automobile safety technology thereafter, as many of the safety devices and functions (Fig. 7.1 [4]) considered at that time are now installed in many vehicles.

7.1.2 Advanced Safety Vehicle (ASV)

Figure 7.2 shows the trends in the dawn of the current ITS-related Japan-US-Europe projects in the 1970s to 1990s.

In Japan, in the late 1980s, with the increase in the number of license holders of the second baby boom generation and the increase in traffic and logistics volume due to the economic boom, projects started in each ministry, and it can be seen that they are connected to the current ITS. Among them, the Ministry of Land, Infrastructure, Transport and Tourism is considering safety technology centered on automobiles, which is the Advanced Safety. This is part of the Advanced Safety Vehicle (ASV) promotion plan. The ASV project started in 1991 and as of 2024, the 7th phase (each phase is a 5-year plan) is underway. The ASV project is being discussed by the ASV Promotion Study Group, which consists of automobile and motorcycle manufacturers (14 companies), scholars, related organizations, and

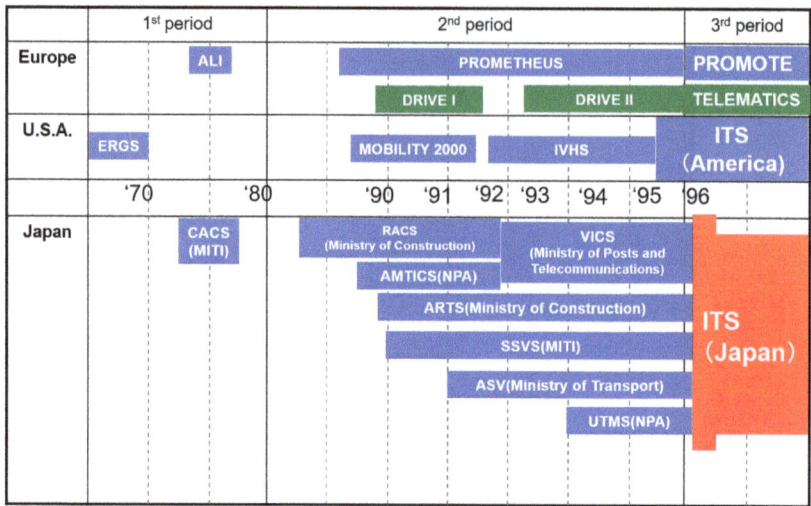

Fig. 7.2 ITS-related projects in the early years (Names of ministries and agencies are as they were at the time)

related government agencies. The overview of each phase of the ASV promotion plan is shown in Table 7.1 as follows.

The concept image in the second phase of ASV shows that the target of preventing accidents and reducing damage caused by collisions, which was mentioned in the previous ESV plan, is realized by applying electronics and control technology that rapidly advanced in the 1980s. At the same time, it also establishes a technical foundation on the vehicle side. Therefore, ASV vehicles are equipped with various sensors and control devices as shown in Fig. 7.3.

In the first phase (1991–1995), technical discussions were initiated for passenger cars, but from the second phase (1996–2000), large vehicles and motorcycles were included in the development targets. A characteristic of the second phase is that this was a period when research on ITS was becoming active worldwide, and ASV was also promoted as part of the "Support for Safe Driving" in the "Overall Concept for Advanced Road Traffic (ITS)" issued in July 1996. In addition, cooperation with road infrastructure was also considered, and in October 2000, joint demonstration experiments on driver acceptance and the appropriateness of road infrastructure facilities were conducted at the Civil Engineering Research Institute of the Ministry of Land, Infrastructure, Transport and Tourism (formerly the Civil Engineering Research Institute of the Ministry of Construction) in Tsukuba City, Ibaraki Prefecture, including cooperation and coordination with the Advanced Highway System (AHS) and seven services (prevention of collision with obstacles ahead, prevention of danger when entering curves, prevention of lane departure, prevention of head-on collision, prevention of right-turn collision, prevention of collision with pedestrians at crosswalks, and maintenance of inter-vehicle distance using road surface information).

7 Technology for Safety Improvement: Automobile Edition

Table 7.1 Overview of the ASV promotion plan

Phase 1: 1991–1995	Phase 2: 1996–2000	Phase 3: 2001–2005
Examination of technical possibilities	Research and development for practical use	Examination for promotion of diffusion
		New technology development
Setting development goals	Formulation of ASV basic philosophy	Formulation of driving support concepts
Verification of accident reduction effects	Formulation of guidelines for ASV technology development	Formulation of ASV diffusion strategy
	Verification of accident reduction effects	Promotion of technology development using communication technology
Phase 4: 2006–2010	**Phase 5: 2011–2015**	**Phase 6: 2016–2020**
Contribution and challenge to accident reduction	Realization of dramatic advancement	Promotion of ASV toward the realization of autonomous driving
Examination and implementation of evaluation methods for traffic accident reduction effects	Formulation of basic plan for driver abnormality response system	Organization of advanced safety technologies with autonomous driving in mind
Formulation of basic design documents for communication-based driving support systems	Formulation of basic design documents for pedestrian-vehicle communication systems	Examination of specific technologies with the aim of setting guidelines for development and practical application
		The spread of autonomous driving technology, including realized ASV technology
Phase 7: 2021–2025		
Further promotion of ASVs for advanced automated driving		
Thorough and effective dissemination strategies for the correct understanding and use of ASV technology, which has become widely used by everyone		
Examination of the appropriate approach to safety technology that prioritizes system operation over driver operation		
Consideration of common specifications for the practical application and dissemination of safety technologies that utilize communications and maps		
Exploration of the scope and level of safety that automated vehicles should possess		

In the third phase (2001–2005), activities were conducted to promote the spread of ASV. In this phase, the "Concept of Driving Assistance" was compiled as a concrete form of the "Basic Philosophy of ASV" formulated in the second phase. For example, in terms of "Ensuring Driver Acceptance," it is stipulated that the driver should be able to confirm the operation of the system. Also, the "Principle of Driver Assistance" clarifies that the driver is the main actor who should drive safely, and ASV technology is positioned as support from the side. Based on this principle, even in the devices that have been put into practical use, the driver can choose the

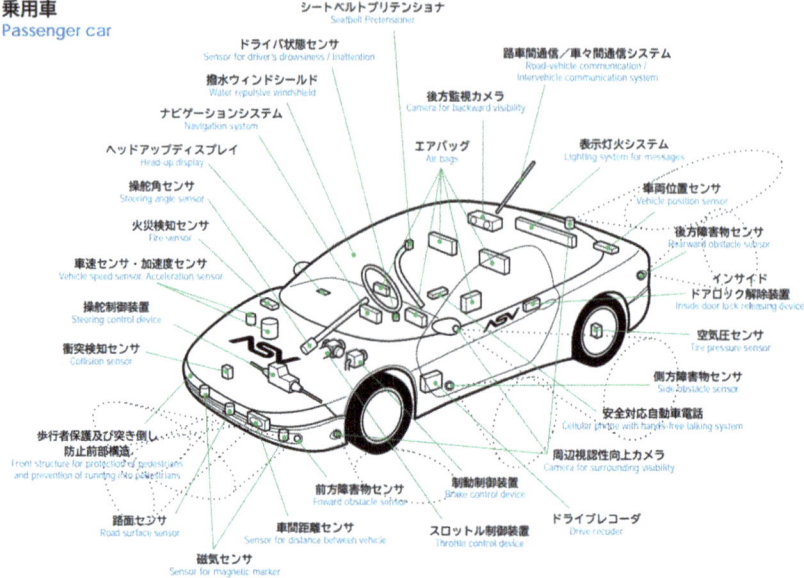

Fig. 7.3 Developing Technologies in the ASV-2 [5]

operation switch of the support system, and the driver can forcibly intervene in the system at the time of operation.

In the fourth phase (2006–2010), basic design was conducted for the practical application of communication-based driving support systems. It is expected that accidents can be avoided by using vehicle-to-vehicle communication (V2V: Vehicle to Vehicle) in situations such as head-on collisions and turns, and support for recognizing surrounding vehicles is also discussed. In addition, in recent years, health-related accidents, which have become a social issue, are also being considered for adaptation to large vehicles by analyzing accidents and considering systems to respond to abnormal driver conditions.

In the fifth term (2011–2015), as part of the study on the dramatic advancement of ASV technology, efforts were made such as the creation of a basic design document for a driver anomaly response system, the creation of design considerations to suppress the occurrence of driver overconfidence, and the clarification of the concept of priority of support when multiple systems operate simultaneously. In addition, as part of the study on promoting the development of communication-based safe driving support systems, the creation of a basic plan for a pedestrian-targeted vehicle-to-pedestrian communication system and the creation of a revised version of the basic design document for a communication-based driving support system considering the next generation were carried out.

In the sixth term (2016–2020), there was a major turning point, and the basic concept of ASV technology, which is the "basic philosophy of ASV" shown in the third term, that ASV technology is based on driver support, was re-examined in light

7 Technology for Safety Improvement: Automobile Edition 125

of the recent activation of the development of autonomous driving technology, and the basic philosophy was summarized into the following eight points under the title "Concept of Driving Support/Autonomous Driving."

1. Operation based on intention
2. Safe driving and stable operation
3. Confirm the content of operation
4. Do not give overconfidence
5. Forced intervention possible
6. Smooth transition
7. Safety does not decline
8. A foundation for acceptance in society is formed

As a result, by assigning the eight concepts to the three parties: the user, including the driver, the system, and society, the position of ASV technology has been clarified not only when the driver is the main operator (Fig. 7.4), but also when the system is the main operator (Fig. 7.5). In line with the assumption of use in autonomous driving, the sixth phase also implemented considerations such as the impact and points to note when introducing autonomous vehicles under mixed traffic, and the evaluation of the accident reduction effect of the autonomous driving system. In addition, with the practical application and popularization of ASV technology, a common definition of ASV technology and a glossary of terms related to

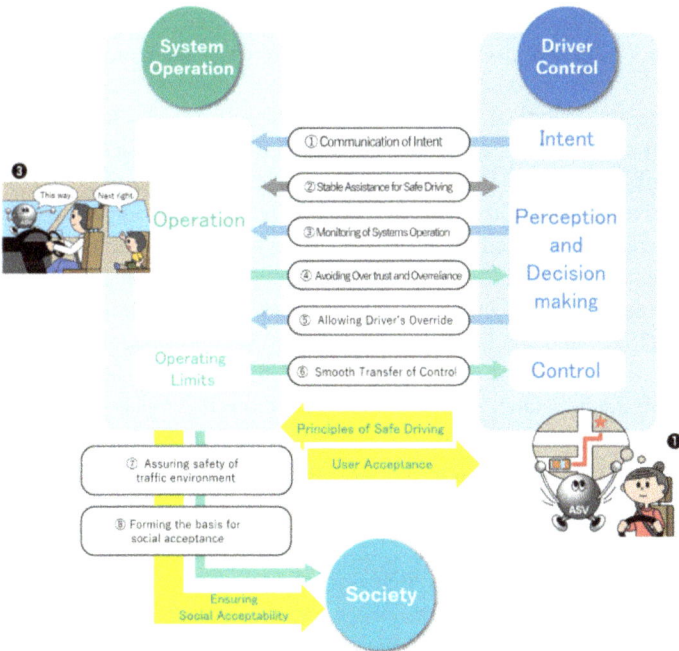

Fig. 7.4 Guiding principles of ASV (when driver-centric driving) [6]

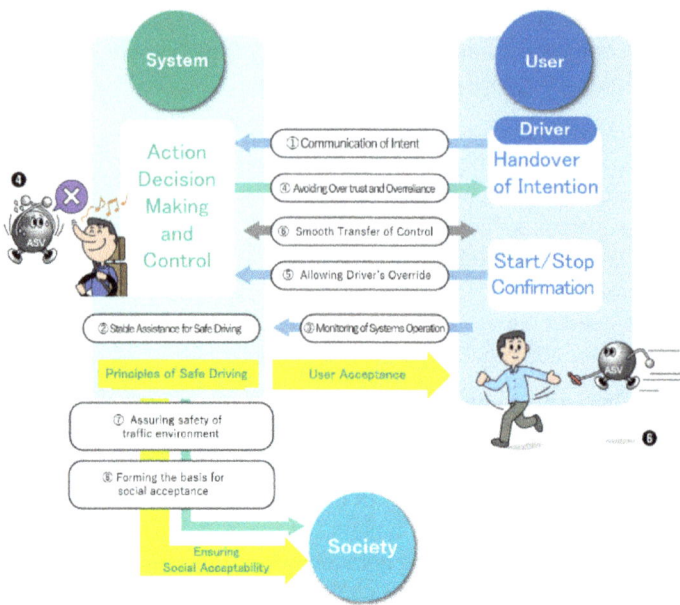

Fig. 7.5 Guiding principles of ASV (When autonomous driving system is primary driver) [6]

autonomous driving, which are often used in newspapers and magazines, were created to avoid confusion due to differences in performance and definitions among manufacturers as part of dissemination and awareness-raising activities for general users [7].

In the specific technical considerations with the aim of setting guidelines for development and practical application, detailed discussions on specific issues are being advanced, such as the consideration of technical requirement issues for advanced driver abnormality response systems such as road side evacuation type in surveillance technology, and the consideration of highway version and general road version separately. In addition, guidelines for systems such as the continuous unmanned platoon driving system by electronic towing, the Intelligent Speed Assistance (ISA), and the last mile autonomous vehicle system have been established, and discussions are continuing toward the practical application of more advanced driving assistance systems and autonomous driving systems [6].

As of 2024, the seventh phase (2021–2025) is underway, and the basic theme is "Further promotion of ASV toward advanced autonomous driving," and the characteristic is that points of consideration are raised on the premise of use in autonomous driving. Some of the items under consideration are shown below.

- The way of safety technology that prioritizes the system's operation over the driver's operation
- Common specifications for the practical application and dissemination of safety technology using communication and maps
- Consideration for exploring the range and level of safety that autonomous vehicles should have

7.1.3 Intelligent Transport Systems (ITS)

As shown in the previous section in Fig. 7.2, automotive safety technology for accident reduction has been positioned as an important field in ITS due to the integration with electronics technology and information and communication technology since the late 1980s. The second ITS World Congress was held in Yokohama in 1995, and in Japan, the voluntary organization Vehicle, Road and Traffic Intelligence Society (VERTIS) was established in 1994 in collaboration with five related ministries (the then National Police Agency, Ministry of International Trade and Industry, Ministry of Transport, Ministry of Posts and Telecommunications, and Ministry of Construction). Later, in 2001, it was renamed ITS Japan, and at that time, the "ITS Basic Strategy Committee" was established, and its role includes functions such as policy recommendation, government-private sector collaboration and coordination, and international strategy, leading to the present.

ITS Japan formulated a mid-term plan over a five-year span for the execution of the missions mentioned earlier. In the first mid-term plan (2006–2010), ITS Japan worked to further popularize the nine fields and 21 services that were put into practical use in accordance with the "Overall Concept for ITS Promotion" created by five ministries in 1996. In addition, in line with the "Guidelines for ITS Promotion" compiled by the Japan ITS Promotion Council in 2004, ITS Japan implemented measures such as the development of a common infrastructure for the realization of the areas expected in the guidelines for ITS promotion, namely, "a safe and secure society," "an environmentally friendly and efficient society," and "a convenient and comfortable society."

In the second ITS mid-term plan (2011–2015), a comprehensive transportation system was proposed with the goal of realizing the long-term ITS vision for 2030, which integrates transportation networks, energy networks, and information and communication networks [8]. Notably, in response to the Great East Japan Earthquake in March 2011, it is highlighted that the plan explicitly mentions the construction of an information infrastructure that can be utilized not only in normal times but also in times of disaster. Furthermore, the fusion of ASV technology and information and communication technology allows for multipop routing through vehicle-to-vehicle communication, sharing information about the road conditions ahead and vehicles in the vicinity, not just in the immediate vicinity of each vehicle. This development toward a system that ensures the safety of the entire group of vehicles is expected to reduce traffic accidents by providing information about the risk of accidents and dangerous events in advance. In addition, the development of support technologies for traffic participants, including pedestrians, bicycles, and new mobility, aims to compensate for the effects of driving proficiency and cognitive decline due to aging, and to realize a transportation society where anyone can move safely [9].

In the third ITS mid-term plan (2016–2020), it broadened its perspective and formulated it as an initiative to support and realize transportation in the future of the country and the city, from the perspective of solving social issues and creating a vibrant society. Therefore, the social issues to be solved and the images of the actual sites to be implemented in society were first organized. The social issues include perspectives related to the current SDGs and well-being and were organized as (1) declining birthrate and aging population, (2) energy and environmental issues, (3) sluggish economic growth, and (4) safety and security. The approach was organized into four categories: (1) breaking down technology discussions from the perspective of specific services and value creation, (2) advancing concrete proposals based on case studies based on field-oriented principles, (3) reconstructing the boundaries of competition and cooperation with new ideas, and (4) building cooperation with a global collaboration in mind. In particular, a major difference from the second phase is that we have typified the targets of our efforts into the following three bases, by comparing the "Grand Design of National Land 2050, [10]" which is the basis of the government's philosophy and approach to national land development looking towards 2050, and the "Recommendations for Future Creation by ITS," in order to show the path to social implementation by assuming specific situations.

- Small Bases: These are approximately 5000 units of population bases with a scale of 10,000 people, bundling together the settlements said to be 65,000 nationwide.
- Advanced Urban Alliance: These are regions where medium-sized cities, which are expected to fall below a population of 300,000 in the future, collaborate and complement each other to form a sphere with a population of about 300,000 necessary for maintaining advanced public services.
- Large Cities: These are cities with a population of over 1 million.

By overlaying the perspectives mentioned earlier on these three types, we can organize the issues hierarchically and visualize what kind of technology is needed to realize the services necessary on the ground. An example of the transportation services layer of this is shown in Fig. 7.6. In reality, a matrix is formed at each layer, and the points of consideration are subdivided. This has solidified the image of realizing detailed comprehensive mobility services under various conditions. Figure 7.7 is an image of the integrated mobility service that ITS aims for in the third mid-term plan, and it is interesting to see that it is similar to the direction currently being discussed in MaaS, which organically combines various means of transportation to realize a safe, secure, and vibrant society. In this context, it is also indicated that convenience is enhanced on demand by utilizing public transportation and sharing appropriately, not by replacing everything with autonomous driving. In the ongoing ITS fourth mid-term plan (2021–2026), an outline has been proposed that aims to evolve toward safer, more secure, comfortable, and efficient mobility through the advancement of the mobility system, and to contribute to the realization of a mobility value chain that supports diverse lifestyles, and the relevance with social implementation of MaaS and autonomous driving is becoming higher.

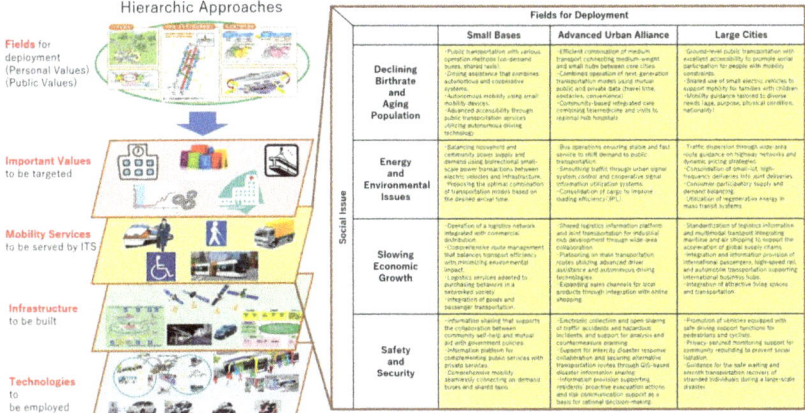

Fig. 7.6 Target matrix in the ITS 3rd mid-term plan (e.g., mobility services layer) [11]

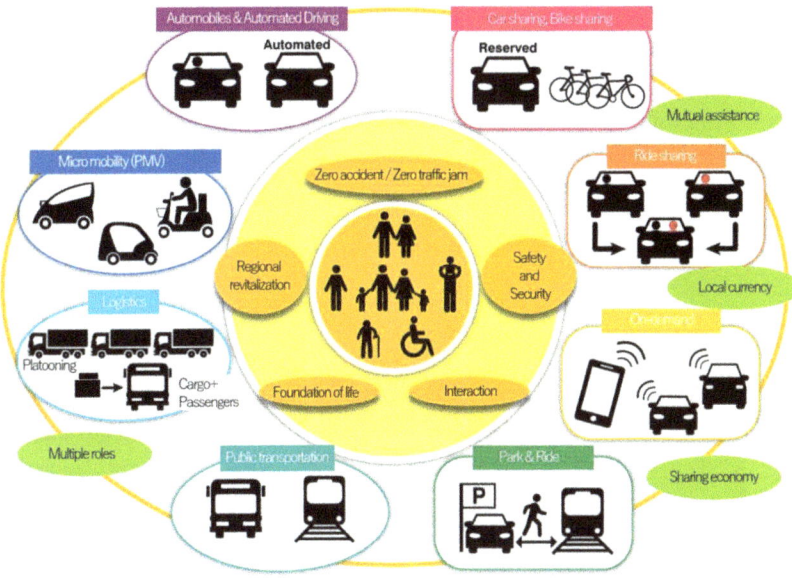

Fig. 7.7 Integrated mobility services diagram in the ITS Japan 3rd mid-term plan [11]

7.1.4 Initiatives Toward the Social Implementation of Autonomous Driving

So far, we have introduced the ESV plan, which has been responsible for advanced technology development as a trend in the development of automobile safety technology, the ASV plan, and the ITS plan to realize safe, secure, and comfortable transportation by incorporating information and communication technology. Among them, the core technology fields known as CASE (Connected, Autonomous, Shared, and Electric) have been attracting particular attention in recent years. This period is often described as a once-in-a-century major reform. In this context, not only competition between companies but also international competition has intensified. Because research and development costs are enormous, it is difficult for a single company to bear them alone. Therefore, national projects are promoting research and development in cooperative areas, while in competitive areas, companies are building relationships such as technical cooperation and technology licensing. Currently, in Japan, discussions are being advanced in the Cross-ministerial Strategic Innovation Promotion Program (SIP) by the Cabinet Office regarding the cooperative area of autonomous driving.

The details of these autonomous driving and driving assistance will be reintroduced in Chap. 8, and from the next section onward, we will introduce other safety improvement technologies.

7.2 Active Safety

The safety improvement technologies shown in Sect. 7.1 will be introduced according to the timing of device operation along the timeline of collision risk occurrence during vehicle driving, as shown in Fig. 7.8.

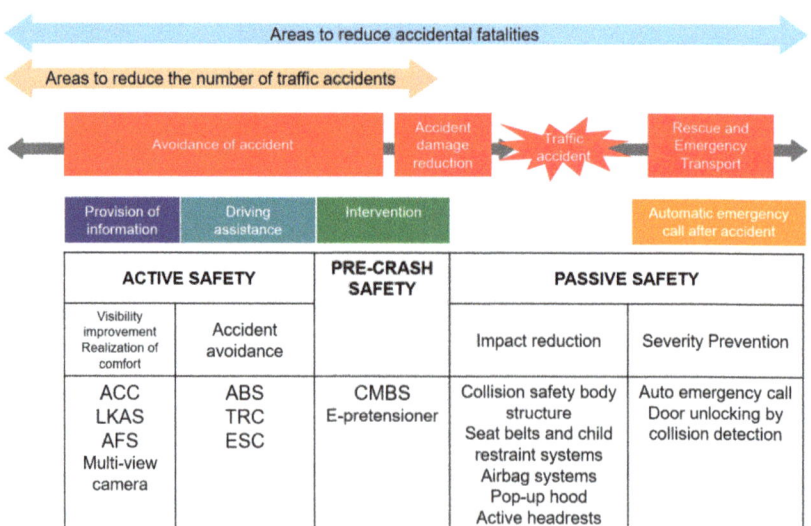

Fig. 7.8 Vehicle Safety Technologies by Accident Prevention and Damage Reduction Stage

7 Technology for Safety Improvement: Automobile Edition

Even in Japan, about 300,000 traffic accidents occur annually in 2023, and the number of accidents has decreased to about half of that 10 years ago, but still more than 350,000 casualties occur annually. Therefore, in order to aim for zero traffic accidents as the ultimate goal, it is important to develop and implement practical use and popularization of active safety technologies that prevent traffic accidents from happening in the first place.

7.2.1 Visibility Improvement Technology

7.2.1.1 Lighting System

High-intensity discharge headlights and LED headlights have been put into practical use to improve the visibility of the driver's forward field of view in dusk, nighttime, and bad weather. In addition, systems such as AFS (Adaptive Front-lighting System), which calculates the vehicle's direction of travel during curve driving from the steering angle and vehicle speed and controls the reflector of the headlight unit to align the illumination direction with the direction of travel, are becoming widespread.

In Europe and other places, daytime lighting to improve the visibility of your vehicle from oncoming vehicles and pedestrians is mandatory, but in Japan, due to the revision of the safety standards of the Road Transport Vehicle Act, auto lights that prohibit no-light driving have been mandatory for new cars from March 2019 (continuous production cars from October 2021), and along with that, DRL (Daytime Running Light) using LEDs has been recognized as daytime running lights. Because LEDs are used, they are often placed around the edge of the headlights in terms of design.

7.2.1.2 Indirect Field of View

Technologies that install multiple cameras on the vehicle body and display the situation of the intersecting lanes at intersections with poor visibility on a monitor, or synthesize the size of the vehicle in the rear image of the vehicle during backing, or use image processing technology to synthesize the around view of the vehicle surroundings and display it on a monitor to improve the driver's understanding and visibility are also in practical use.

In addition, there are also night vision systems that illuminate the front with near-infrared light at night and visualize and emphasize pedestrians and road conditions that are difficult to see with the naked eye with a near-infrared camera to alert the driver (Fig. 7.9).

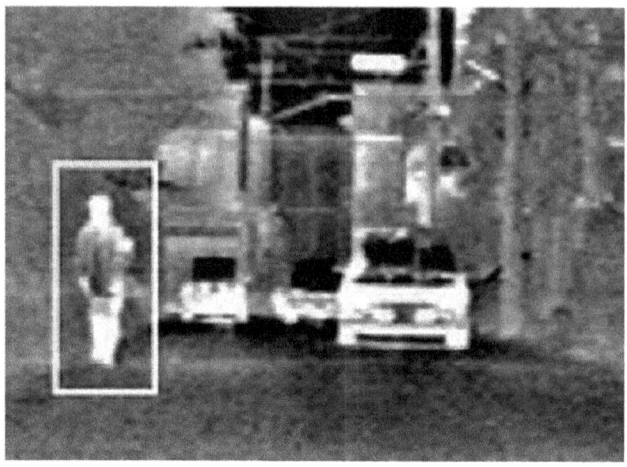

Fig. 7.9 Example of pedestrian highlighting by intelligent night vision system. ((Provided by Honda Motor Co., Ltd.) [12])

7.2.2 Driving Work Load Reduction Technology

7.2.2.1 Speed Control

Traditional cruise control was a function to maintain a constant speed on highways and the like, but now, using millimeter-wave radar and stereo cameras, vehicles can recognize the situation ahead and automatically follow the speed while maintaining the distance from the vehicle ahead. This is a constant speed driving/distance control device (Adaptive Cruise Control, ACC) that is not only used in high-speed areas such as highways but also in traffic jams on general roads and is also practical as a low-speed ACC aimed at warning of approaching vehicles ahead and preventing rear-end collisions. In particular, low-speed ACC also serves as a collision mitigation brake system (CMBS) and is considered effective in reducing the number of accidents in areas with many accidents, such as near intersections.

In Japan, since November 2021, new domestic cars (continuously produced cars are December 2025) are required to be equipped with an Advanced Emergency Braking System (AEBS), and manual transmission cars are also equipped with throttle rotation control and AEBS. Therefore, by partially utilizing its sensors and control system, even manual transmission cars are equipped with a simple ACC, which also reduces the driving load during long-distance driving on highways.

This throttle rotation control also operates as a blipping function that increases the rotation speed during the shift down of the manual transmission, making it easier to synchronize and supporting the driver's smooth shifting operation.

*CMBS and AEBS are described differently because their operating conditions are different.

7.2.2.2 Head-Up Display (HUD)

A HUD is a system that projects necessary information for driving onto a reflective LCD panel or holographic optical element placed in front of the windshield or meter hood as if focusing at infinity. Because it allows the driver to read information without looking down at the instrument panel, it is expected to prevent distracted driving and reduce fatigue.

Also, by projecting information to overlap with the driver's forward field of view, effects such as 3D navigation can be achieved, creating an augmented reality (AR) effect.

7.2.2.3 Driver Fatigue Detection System

This system detects the driver's fatigue and attention levels from the fluctuations in steering operation and facial analysis by the in-car camera. When fatigue is detected, it issues a warning to prevent accidents. For example, there are systems that monitor the direction of the driver's face and the state of the eyes while driving, detect lack of attention, and learn the driving pattern for the first 15 minutes of driving, detecting a decrease in attention from steering operations and other factors using fuzzy theory. In all cases, when a decrease in attention is detected, a warning is given by an alarm sound or a warning display. In addition, for swaying associated with decreased attention, the system alerts the driver with a lane departure warning, and, if the warning continues, it can be expected to prevent accidents by advancing the timing of the collision damage mitigation brake support [13].

7.2.2.4 Active Speaker System

Recent passenger cars have improved body rigidity and excellent soundproofing performance, improving the quietness inside the car to enhance the comfort of the passengers and reduce the driver's fatigue. On the other hand, due to electrification and noise regulations, engine sounds that provide information about the driving state of the vehicle are no longer audible. Therefore, by synthesizing engine sounds from the degree of accelerator opening and playing the synthesized sounds from the speakers inside the car, it is easier to imagine the driving state, and in some sports cars, it enhances the exhilaration of the driver's operation. In addition, for external sounds, the system uses a noise-canceling function to cancel out as much noise as possible and emphasizes the direction of incoming sounds, such as the presence of a rear vehicle or the siren of an emergency vehicle, to alert the driver.

7.2.3 Vehicle Motion Performance Improvement Technology

7.2.3.1 Electronic Stability Control (ESC)

Even if the driver turns the steering wheel, if the wheels are not rotating, no lateral force is generated to turn the vehicle. Therefore, current vehicles are almost standard equipped with ABS (Anti-lock Brake System). ABS constantly monitors the wheel rotation state, and when the wheels start to lock during braking on slippery surfaces, it instantly releases the brake pressure to eliminate wheel lock and then increases the brake pressure to a level that does not lock, ensuring braking performance while enabling steering operation. Currently, TCS (Traction Control System), which prevents the wheels from spinning during start-up and acceleration, and ABS are linked, and by actively integrating the controlling the tire force generated by each wheel, the vehicle's swaying and lateral sliding are prevented and reduced. ESC (Electronic Stability Control) has been mandatory for new cars since October 2012. Furthermore, integrated control with electric power steering EPS and others is progressing, and by improving the total vehicle driving performance with these, the realization of danger and accident avoidance in case of emergency is being pursued.

7.2.3.2 Tire Pressure Monitoring System (TPMS)

For autonomous driving and driver assistance systems to achieve ideal vehicle behavior, it is necessary for the condition of the tires to be appropriate. In particular, if the air pressure is not at the correct value, not only the braking and driving force but also the cornering force for bending the vehicle decreases. For example, if the rear wheels become like a slow puncture, the balance of the forces generated from the front and rear tires is disrupted, and even if ESC is installed, appropriate control cannot be achieved, which can induce a tendency to spin. Therefore, it is a system that monitors whether each wheel's tire is not punctured or the air pressure is not abnormal through pressure sensors built into each wheel and warns the driver when the air pressure decreases. Maintaining appropriate air pressure not only improves safety but also improves fuel efficiency. It is becoming mandatory in the United States and other major countries in Europe, but in Japan, it is limited to being mandatory for run-flat tires, which are difficult to understand the puncture situation of, and is currently under consideration for legislation for regular tires.

7.3 Passive Safety

In Sect. 7.2, we discussed technologies for improving safety to prevent accidents in advance, but it is difficult to completely prevent accidents with them, and it is also important to improve the performance of passive safety to minimize injuries to

7 Technology for Safety Improvement: Automobile Edition

people in the event of a collision accident. Here, we introduce representative technologies for that.

7.3.1 Collision Damage Reduction Technology

7.3.1.1 Crushable Body

In a collision accident, the momentum due to the vehicle's speed affects the size of the impact transmitted to the driver. If the body is very rigid, the impact is transmitted to the driver all at once, resulting in a high damage value for the driver. Therefore, the impact force transmitted is reduced by absorbing energy through the deformation of the frame in the engine room part called the crushable zone at the time of collision. Currently, the amount of frame deformation over time against impact is calculated in advance at the design stage, and the frame is designed so that the maximum value of the impact is below the dangerous value for the driver.

7.3.1.2 Compatibility

In a collision accident between vehicles under mixed traffic, the function of the aforementioned crushable zone is halved if the frame positions of the vehicles collide differently due to the difference in vehicle size. Therefore, even when vehicles of different sizes, such as light vehicles and ordinary vehicles, collide, the body structure is designed so that they impact each other's frames, ensuring compatibility (Fig. 7.10).

7.3.1.3 Pre-tensioner ELR Seatbelt with Load Limiter

As a technology to mitigate the impact on passengers during a collision, a pre-tensioner ELR seatbelt with a load limiter is used. This seatbelt instantly retracts the slack in the seatbelt when a collision is detected, and then feeds out the belt when a certain load is applied, reducing the burden on the chest.

Fig. 7.10 Compatibility body structure. ((Provided by Honda Motor Co., Ltd.) [14])

7.3.1.4 SRS Airbag System

The SRS airbag system is now commonly installed not only in the driver's seat but also in the passenger seat. Furthermore, it detects a collision in about 0.015 s and changes the deployment speed of the airbag according to the collision speed or continuously changes the deployment volume, improving the protective performance for passengers of different sizes. Also, the side airbag system for side collisions, which deploys in a curtain-like manner to effectively protect the head and neck, has been put into practical use.

7.3.2 Pedestrian Damage Mitigation Technology

7.3.2.1 Body Design Shape

As a technology to reduce the damage to pedestrians in the event of a collision, a structure that absorbs impact by ensuring space in the bumper, fender, and inside the bonnet in the body design and making it easy to dent is adopted. In addition, by eliminating as much as possible the protruding parts of the body tip, the snagging of the pedestrian's body at the time of collision is eliminated. Therefore, retractable headlights that become convex when deployed are no longer adopted.

7.3.2.2 Bonnet Hood Structure

The hinge part of the bonnet itself is designed to bend and absorb impact when a strong impact is applied. In addition, a pop-up style bonnet hood that pops up the bonnet hood the moment a collision is detected to increase space is also adopted.

7.3.2.3 Pedestrian Airbag System

When a collision with a pedestrian occurs from the front of a passenger car, if the vehicle speed is high and the pedestrian is an adult, the person's legs first contact the front bumper of the vehicle, and the momentum causes the upper body including the head to fall violently towards the bonnet. In that case, the pedestrian's head may suffer fatal injuries by hitting the front window glass or A-pillar beyond the bonnet hood. Therefore, when the vehicle detects a certain collision speed, it sends an ignition signal to the pedestrian airbag system stored between the bonnet hood and the front window at the same time as the collision is detected at the bumper part and deploys the airbag to protect the pedestrian's head from impact. This system has also been put into practical use.

On the other hand, once a collision occurs, when it becomes necessary to replace parts such as bumpers and windshields where sensors are installed, not only sensor replacement but also calibration work for the sensors and the system is required. This work can only be done at designated factories capable of performing the manufacturer-specified advanced work, so the repair unit price of recent commercial vehicles has been on the rise, and despite the absolute number of accidents decreasing, it is also cited as a factor in the increase in voluntary insurance premiums.

According to the Traffic Safety White Paper, more than half of the legal violations by the first party in traffic fatalities in Japan are violations of the duty of safe driving, and it is said that the majority of traffic accidents as a whole have been caused by drivers, so it is expected that accident reduction can be achieved through further functional improvements of advanced driver assistance systems at the active safety stage.

7.4 Other Safety Technologies

7.4.1 Safety Improvement Technology for Super-Aged Societies

Japan, with a high aging rate of 28.4% as of 2019, which is the proportion of people aged 65 and over in the total population, is notably advancing into a super-aged society even among the world. In addition, since the motorization that began in Japan in the 1950s, the generation of drivers who obtained their driver's licenses are approaching the late elderly (75 years and older), and traffic accidents in which the elderly are injured or killed are becoming a major social issue, while accidents involving cars driven by the elderly are also increasing. Among these, there are cases caused by dementia, and currently, drivers aged 75 and over are required to undergo cognitive function tests when renewing their licenses due to legal revisions. Also, because everyone experiences a decline in judgment and physical abilities with age, the following safety technologies have been developed. In addition, efforts are being made to promote awareness of safety support cars for the elderly through public-private partnerships.

7.4.1.1 Pedal Misapplication Acceleration Suppression Device

In fatal accidents involving elderly drivers aged 75 and over, 30% are due to inappropriate operation (14.6% for inappropriate steering operation, 7.3% for pedal misapplication), and the proportion is significantly larger compared to those under 75 [15].

Therefore, it is a function that is expected to reduce these accidents by suppressing erroneous starts and warning the driver of the danger of collision with warning sounds or meter displays, etc., when the accelerator pedal is strongly pressed despite obstacles in front or behind. The collision damage mitigation brake (automatic brake), and this sudden acceleration suppression device are the core safety improvement technologies in the safety support car set up to suppress accidents of the elderly.

7.4.1.2 Vehicle Rear Confirmation Device When Reversing

This is a system installed at the rear of a car that detects obstacles or objects behind the vehicle when the driver is reversing and provides a warning. This device typically uses a small camera mounted on the car's bumper or rear hatch to display the rear view, providing visual information to the driver through a monitor. In particular, it is effective in avoiding contact with low-height children who are crouched in close proximity to the vehicle, which is difficult to confirm with rear indirect vision such as side mirrors and room mirrors when reversing. These accidents are considered important internationally, and in addition to the back camera type, methods to confirm the presence or absence of obstacles by sensors such as ultrasonic sonar have been adopted at the UN WP29, and in Japan, it has been made mandatory for new cars (continuous production cars after May 2024) after May 2022.

7.4.2 New Safety Improvement Technology Using Connected Technology

Connected technology, which equips vehicles with communication functions and performs data transmission and reception, is widely installed and used in current production vehicles by incorporating a SIM into the vehicle. In this session, we will describe relatively new safety improvement technologies using connected technology.

7.4.2.1 One-Touch Reporting (Such as QQ Call)

In the event of an accident or if a passenger becomes ill, just pressing the help button will connect to a dedicated operator and send the vehicle's location information, making the time until the emergency vehicle is dispatched shorter. Also, depending on the vehicle type, there is a function to automatically report when the airbag is activated, so even if you lose consciousness in an accident, a report will be made, which can shorten the arrival time of the emergency vehicle.

7.4.2.2 Intelligent Speed Assistance (ISA)

ISA is a system that recognizes the speed limit of the road in real time and warns the driver and restricts the speed of the vehicle as necessary. Its mechanism uses sensors such as cameras and GPS to recognize the speed limit of the location where the vehicle is currently located. It recognizes road signs on the road by video and accurately identifies the speed limit by accessing the database on the network server using connected technology. Then, if ISA determines that there is a possibility of exceeding the speed limit, it displays the speed limit on the display and implements a voice warning, and at the same time, restricts the speed of the vehicle through throttle operation or cruise control as necessary, assisting the driver to comply with the speed limit. ISA is being considered for legal obligation in some regions, such as the EU, as an effective system to assist in speed control for vehicles that use residential roads with many accidents with pedestrians and bicycles where Zone 30 is set as a detour.

Furthermore, a similar system to this feature is the speed limiter release function. This is installed in some sports cars, and it uses GPS to determine the current location. When the vehicle's presence is confirmed within a preset dedicated circuit course, the speed limiter is released, allowing the vehicle's performance to be maximized in sports driving or racing.

References

1. Okami Y (1976) Safety and ESV plan. IATSS Review 2(2)
2. Toyota Motor Corporation, "Toyota Motor Corporation 75-year history – Part 2, Chapter 2, Section 2, Item 4: Development of Toyota ESV." http://www.toyota.co.jp/jpn/company/history/75years/text/index.html
3. Sano S (1979) RSV plan and prospects for safety research. IATSS Review 5(4)
4. Society of Automotive Engineers of Japan (2009) Encyclopedia of Automobiles. Maruzen, p 297
5. Ministry of Land, Infrastructure, Transport and Tourism (2002) 3rd ASV promotion plan pamphlet. https://www.mlit.go.jp/jidosha/anzen/01asv/data/asv3pamphlet.pdf
6. Ministry of Land, Infrastructure, Transport and Tourism (2021) Report on the Results of the 6th Advanced Safety Vehicle (ASV) Promotion Plan. https://www.mlit.go.jp/jidosha/anzen/01asv/report06/index.html#philosophy
7. Ministry of Land, Infrastructure, Transport and Tourism (2021) Overview of autonomous driving related terms frequently used in newspapers, magazines, etc. https://www.mlit.go.jp/jidosha/anzen/01asv/report06/file/siryohen_4_jidountenyogo.pdf
8. ITS Japan, "2nd Mid-Term ITS Plan (2011–2015)." https://www.its-jp.org/katsudou/chukei/id210_1/
9. ITS Japan (2013) Proposals for creating the future with ITS – realizing a society where anyone can move comfortably anywhere. http://www.its-jp.org/document/20131017/ITS-future-vision_j_131010.pdf
10. Ministry of Land, Infrastructure, Transport and Tourism (2014) Grand Design of National Land 2050. https://www.mlit.go.jp/kokudoseisaku/kokudoseisaku_tk3_000043.html
11. ITS Japan, ITS Annual Report 2018 Chapter2, 2018.
12. Honda R&D Co., Ltd. (2004) World's first 'intelligent night vision system' that detects pedestrians and alerts the driver – to be equipped in the legend to be released this autumn as a nighttime driving support system. http://www.honda.co.jp/news/2004/4040824a.html
13. Mitsubishi Fuso Truck and Bus Corporation (2008) Driver Attention Monitor (MDAS-III) Wins 'Technology Development Award' at the 58th Society of Automotive Engineers of Japan Awards. http://www.mitsubishi-fuso.com/jp/news/news_content/080424/080424.html
14. Honda R&D Co., Ltd. (2003) Compatibility Body. http://www.honda.co.jp/tech/auto/compatibility/
15. Regarding fatal accidents of the elderly, Safety Support Car homepage. https://www.safety-support-car.go.jp/analysis/

Recommended Literature

16. Ministry of Land, Infrastructure, Transport and Tourism, Advanced Safety Vehicle Promotion Study Group (2008) Research report on advanced safety vehicles – ASV (Advanced Safety Vehicle) research results and future technical guidelines. https://www.mlit.go.jp/jidosha/anzen/01asv/data/asv1report.pdf
17. Ministry of Land, Infrastructure, Transport and Tourism, Advanced Safety Vehicle Promotion Study Group (2001) Report on the Advanced Safety Vehicle (ASV) Promotion Plan (Phase 2). https://www.mlit.go.jp/jidosha/anzen/01asv/data/asv2report.pdf
18. Ministry of Land, Infrastructure, Transport and Tourism, Advanced Safety Vehicle Promotion Study Group (2006) Report on the Advanced Safety Vehicle (ASV) Promotion Plan – About the Results of the 3rd Phase ASV Plan. https://www.mlit.go.jp/jidosha/anzen/01asv/data/asv3seikahoukokusyocorrection.pdf
19. Ministry of Land, Infrastructure, Transport and Tourism, Advanced Safety Vehicle Promotion Study Group (2011) Report on the Advanced Safety Vehicle (ASV) Promotion Plan – About the Results of the 4th Phase ASV Plan. https://www.mlit.go.jp/jidosha/anzen/01asv/data/asv4pamphlet_seika.pdf
20. Ministry of Land, Infrastructure, Transport and Tourism, Advanced Safety Vehicle Promotion Study Group (2016), Report on the Advanced Safety Vehicle (ASV) Promotion Plan – About the Results of the 5th Phase ASV Plan. https://www.mlit.go.jp/jidosha/anzen/01asv/data/asv5report.pdf
21. Ministry of Land, Infrastructure, Transport and Tourism, Advanced Safety Vehicle Promotion Study Group (2021) Report on the Advanced Safety Vehicle (ASV) Promotion Plan – About the Results of the 6th Phase ASV Plan. https://www.mlit.go.jp/jidosha/anzen/01asv/report06/file/asv6_houkokusho_honpen.pdf

Open Access This chapter is licensed under the terms of the Creative Commons Attribution-NonCommercial-NoDerivatives 4.0 International License (http://creativecommons.org/licenses/by-nc-nd/4.0/), which permits any noncommercial use, sharing, distribution and reproduction in any medium or format, as long as you give appropriate credit to the original author(s) and the source, provide a link to the Creative Commons license and indicate if you modified the licensed material. You do not have permission under this license to share adapted material derived from this chapter or parts of it.

The images or other third party material in this chapter are included in the chapter's Creative Commons license, unless indicated otherwise in a credit line to the material. If material is not included in the chapter's Creative Commons license and your intended use is not permitted by statutory regulation or exceeds the permitted use, you will need to obtain permission directly from the copyright holder.

Chapter 8
Automated Driving and Driver Assistance

Yoichi Sugimoto and Kunimichi Hatano

Automated driving is expected to be utilized for reducing the number of traffic collisions, providing transportation with more flexibility, solving the problem of driver shortages in logistics, and realizing smoother, more comfortable, and greener traffic flow. However, besides technical challenges, there are also issues of safety, economic viability, and societal acceptance, which make it hard to claim that widespread adoption of automated driving is progressing all at once. This chapter will provide an overview of the current situation of automated driving technology and driver-assistance technology, which is a use of automated driving technology and likely to become more common soon.

8.1 History and Overview of Automated Driving

8.1.1 History of Automated Driving

The Japanese term "Jidosha" is derived from the French word "automobile," which means "self-moving." However, throughout the approximately 250-year history of the automobile, it has been driven by a human. The idea of a car driving itself, the original meaning of "automobile," has been a long-held dream of humanity. A ride attraction where you can view Futurama, a model of a future city that includes

---

Y. Sugimoto (✉)
Innovative Research Excellence, Honda R&D Co., Ltd., Haga, Japan
e-mail: ysugimo@icloud.com

K. Hatano
SDV Business Development Unit, Automobile Operations, Honda Motor Co., Ltd.,
Haga, Japan
e-mail: kunimichi_hatano@jp.honda

© The Author(s) 2026
Pioneering the Future for Traffic and Safety Sciences,
https://doi.org/10.1007/978-981-96-0676-4_8

Fig. 8.1 Driverless vehicle concept [1]

automated highways, was exhibited at the New York World's Fair in 1939. Figure 8.1 is a newspaper advertisement from an electric company in 1956, showing passengers in a driverless vehicle enjoying a game, with the slogan "Electricity may be the driver" [1].

One of the major objectives of the automotive industry in developing automated driving technology is to reduce traffic collisions. Most traffic accidents are caused by human errors. If human errors can be eliminated through automated driving technology, a significant reduction in traffic collisions can be expected.

Since around 1970, ESV (Experimental Safety Vehicle program) started internationally [2]. This project was aimed at dramatically improving the safety of automobiles by researching and developing various safety technologies and demonstrating them with experimental vehicles. Many efforts were made to improve safety, such as passive safety performance and vehicle stability performance. Initially, there was no automated driving in the ESV, but Honda, which participated in the ESV project from 1971, developed a driverless driving system to safely perform the J-turn test for steering stability. The J-turn test is a sudden turning test where the steering angle is input 180 degrees at once from a straight run at 110 km/h in the shape of a J, which was a performance limit test for vehicles at that time. Although this experimental vehicle was unmanned, it was not automated driving because it was controlled by remote control, but it can be said that the control technology for the system to operate the vehicle was being worked on from that time.

Subsequently, various projects were promoted to realize automated driving. In the 1990s, representative ones were PATH in California, USA, and AHS in Japan [3, 4]. The technologies developed in these projects were not for cars to independently drive themselves but required road-side environment preparation. Specifically, it was a system that controls a trajectory of the vehicle while detecting magnetic nails embedded in roads with onboard sensors. In Japan's AHS, demonstration experiments were conducted on the Joshinetsu Expressway in 1996 before it opened, but it did not lead to practical use because it required investment in infrastructure and subsequent facility maintenance and other operating costs.

One of the triggers to accelerate development of more independent automated driving technology was the Grand Challenge by DARPA (Defense Advanced Research Projects Agency, a research institution of the US Department of Defense) [5]. It was a race where automated driving vehicles run unmanned on a course of about 240 km set in the desert outskirts of California within 10 hours, and the prize for the winner was $1 M. The first time in 2004, there were no teams that completed the race, but in the second time in 2005, five teams completed the race, and Stanford University won in 6 hours and 54 minutes.

In 2007, the Urban Challenge was held on an urban course set on the site of a former air force base [6]. This race was based on a more realistic automated driving scenario where automated vehicles execute missions such as passing through intersections safely while complying with traffic laws and mixed driving with other vehicles driven by human drivers.

The automated driving technology developed through the Grand Challenge and Urban Challenge is being applied to current systems.

8.1.2 Overview of Automated Driving Technology

A driver performs dynamic tasks related to driving a car, such as recognizing surrounding road environment, making necessary judgments while predicting risks, and operating a car. In automated driving, the system performs all dynamic driving tasks, eliminating a need for a driver to be involved. Currently, social implementation of automated driving is still limited, but systems that support a driver's dynamic driving tasks are beginning to spread widely as applications of automated driving technology. This driver assistance system, is one in which a human and a system drive in parallel and cooperate. The system also has sensors to perceive surrounding road environment, predicts risks, makes judgments, and outputs control instructions. It provides information to the driver as needed, alerts them, and issues a warning when a hazard is imminent. It assists in operations to reduce driving burden on the driver and intervenes with brakes to assist in collision avoidance and damage reduction when a collision is imminent.

Figure 8.2 shows a more concrete example of a configuration of an automated driving system. The vehicle is equipped with cameras, millimeter-wave radars, and LiDAR (light detection and ranging) s as external sensors to recognize its surroundings. A millimeter-wave radar emits millimeter-wave radio waves around the vehicle and receives radio waves reflected by target objects to measure relative distances from time delay and relative speeds from frequency shift due to the Doppler shift. A LiDAR scans surroundings of the vehicle with a high-resolution infrared beam and measures relative distances from time it takes for the light to travel back and forth. In addition, the vehicle's position related to the map is identified using the Global Navigation Satellite System (GNSS). The recognition algorithm uses information from the external sensors to perceive relative positions and speeds of objects around it, shape of lanes and roads, and road environment information such as traffic signs.

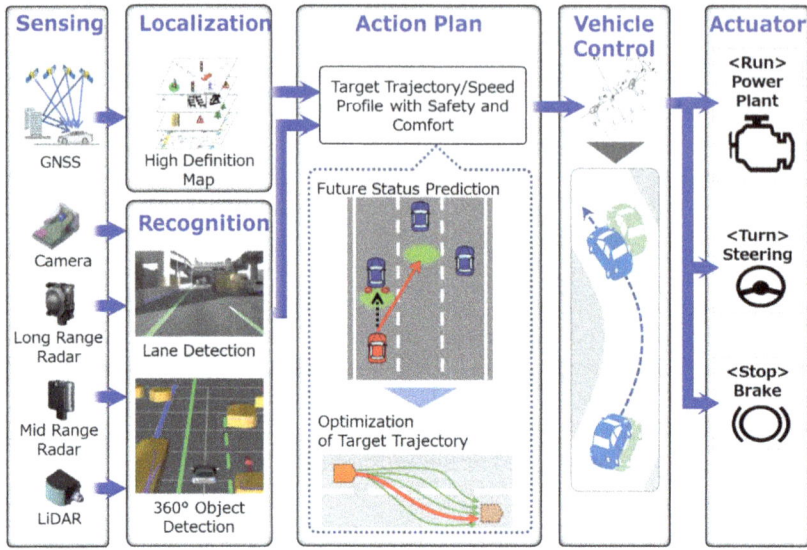

Fig. 8.2 A configuration example of an automated driving system

Based on this information, it also predicts future states, generates a behavior plan that can achieve smooth and comfortable behavior while ensuring safety, and determines the target trajectory and speed profile for the vehicle. The vehicle is controlled by outputting instructions to the power plant, steering actuator, and brake actuator so that the vehicle follows these target values.

8.2 Definition of Driving Automation Levels and Practical Use of Driving Automation Systems

8.2.1 Level Definition by SAE

There are various levels of automated driving systems and driver assistance systems, and the "Taxonomy and Definitions for Terms Related to Driving Automation Systems for On-Road Motor Vehicles" by the Society of Automotive Engineers (SAE) is widely referenced [7, 8] as the definition of levels. Also in Japan, the SAE definition has been adopted since 2017 [9]. This definition is often referred to as the level of automated driving, but it is more accurately the level of driving automation systems, and not all levels are automated driving systems.

Before explaining the level definition, let's explain some important terms.

Dynamic Driving Task (DDT): Real-time operational and tactical functions necessary to operate a vehicle. The OEDR (Object and Event Detection and Response) in the figure refers to monitoring the surrounding traffic environment, detecting objects and events, and executing appropriate responses to them. It does not include strategic functions such as selecting routes or waypoints.

Driving Automation: The performance of part or all of the DDT.

Operational Design Domain (ODD): The operating conditions under which a given driving automation system is designed to function. This includes conditions such as environment, geography, time of day, traffic, roadways, etc. It is also referred to as a limited domain.

Minimal Risk Condition (MRC): A stable and stopped condition brought about by a user or an Automated Driving System (ADS) to reduce the risk of a crash when operation cannot be continued.

Fallback: The response by a user to either perform the DDT or achieve a MRC, or the response by an ADS to perform the DDT after occurrence of a system failure or upon an ODD exit.

Request to Intervene (RTI): An alert from a level 3 ADS to a user indicating that s/he should promptly perform the fallback.

The scope of this level definition is systems that control driving automation on a sustained basis, and active safety systems that support to avoid collisions, such as collision damage mitigation braking or Electronic Stability Control (ESC), which only intervene when a collision is imminent, are not included. However, active safety systems may be installed in driving automation systems at any level, and specifically for ADS, collision avoidance and mitigation functions are part of automated driving system functionality.

There are six levels from 0 to 5.

Level 0 is where a driver performs the entire DDT.

Levels 1 and 2 are driver support, which executes vehicle motion control as part of the Dynamic Driving Task (DDT). Level 1 executes vehicle motion control in either the longitudinal or lateral direction, while level 2 executes vehicle motion control in both the longitudinal and lateral directions. The driver must perform Object and Event Detection and Response (OEDR), i.e., continuously monitor the surrounding traffic environment and execute appropriate responses. Also, the driver needs to monitor the system. The authority of the DDT is the driver.

Levels 3, 4, and 5 are automated driving, which performs the entire DDT and becomes the authority of the DDT while it is functioning.

Level 3 is called "conditional driving automation," where the system performs the entire DDT within a limited Operational Design Domain (ODD). However, the user is expected to execute an appropriate fallback in response to a Request to Intervene (RTI) from the system.

Level 4 is called "high automation," where the system performs all DDTs within a limited ODD. When a fallback is necessary, the system itself executes it to achieve the Minimal Risk Condition (MRC).

Level 5 is called "full automation," where there are no limitations due to the ODD, and the system performs all DDTs in all road environments where a human driver can drive.

8.2.2 Practical Use of Automated Driving Systems

Although the levels of driving automation systems have been explained, a higher level does not necessarily mean higher technical difficulty. Indeed, level 5, which allows for automated driving anytime and anywhere, has extremely high technical difficulty. However, for level 4 and below, it depends on the breadth of its ODD.

Figure 8.3 shows the level of driving automation on the vertical axis and the breadth of the ODD on the horizontal axis.

There are two approaches toward the ideal of level 5 full automation. One is the approach for personal cars, which have traditionally been sold or leased, and the other is the approach for new mobility services.

Personal cars need to be usable anywhere in Japan if they are sold in Japan. Starting from nationwide motorways, efforts are being made to expand the road conditions and driving conditions where they can be used. In 2021, level 3 automated driving during traffic congestion on motorways was put into practical use for the first time in the world in Japan.

On the other hand, there is a demand for new mobility services to address driver shortages and expectations for labor cost reduction, and early deployment of level 4, which eliminates the need for a driver, is desired. The technical difficulty of practical use of level 4 is high, so it is being approached by limiting its ODD narrowly (for example, driving at low speed on a predetermined route with electromagnetic guide lines). In Eiheiji Town, Fukui Prefecture in Japan, after actual operations had been made with level 3 automated driving with remote monitoring, an automated driving service at level 4 started in May 2023 [10]. Also in the United States and China, services are being introduced while limiting the operating area.

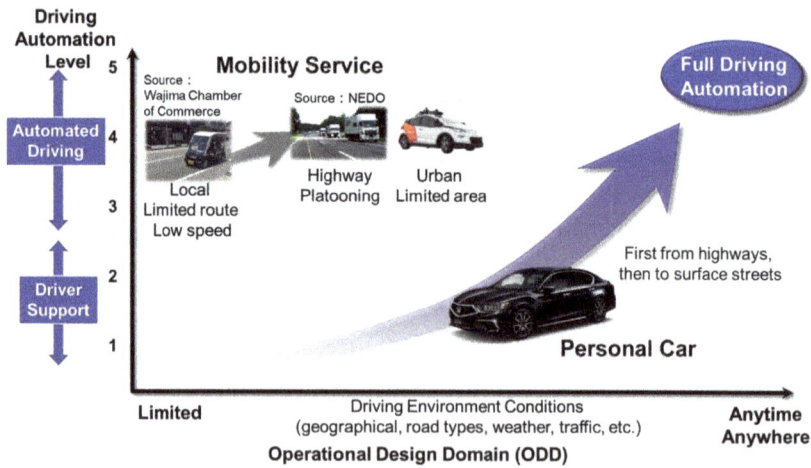

Fig. 8.3 Driving automation level and operational design domain (ODD)

8 Automated Driving and Driver Assistance

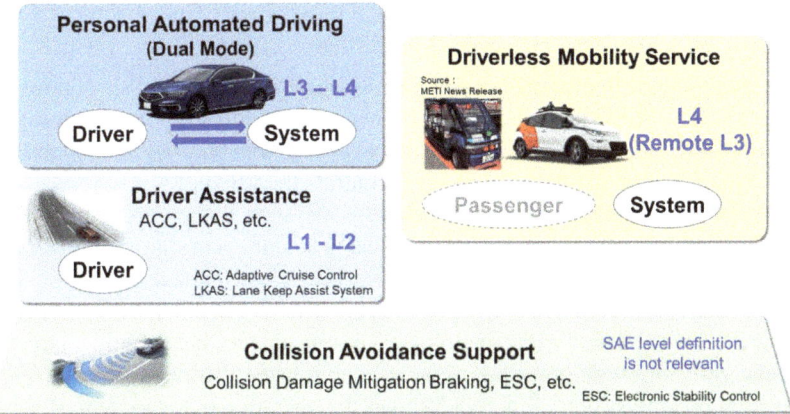

Fig. 8.4 Categories of driver assistance and automated driving

It is difficult to organize driving automation systems only by level. In Fig. 8.4, the currently deployed driver assistance systems and automated driving systems have been organized as categories.

Firstly, as a base system, there is collision avoidance support that assists in avoiding traffic collisions. This category includes such systems as collision damage mitigation braking and ESC, which have significant effects on traffic collision reduction and damage mitigation. These systems only intervene when a collision is predicted to be imminent, so they are not included in the SAE level definition.

On the left part above collision avoidance support in Fig. 8.4 is the area of personal cars, and first there are driver assistance systems from level 1 to 2. Typical functions include Adaptive Cruise Control (ACC), which controls the longitudinal directions, and the Lane Keep Assist system (LKA, also called Lane Centering Assist), which controls the lateral direction. These are driver assistance systems, so the driver needs to monitor the surroundings for safety as usual and perform DDT as needed.

Above that is the automated driving system of levels 3 or 4 for personal cars, but it is not level 5 where automated driving is possible anytime and anywhere, so a dual-mode vehicle that switches between the driver performing DDT and the system performing DDT is assumed.

Further to the right is the area of mobility services, where automated operation without a driver is possible. The human onboard is no longer a driver but a passenger, and the system becomes the authority of DDT. Level 3, which supports fallback by remote communication, could also be considered, but basically, level 4 is aimed at.

These categories will be explained in order from the next section.

8.3 Collision Avoidance Support

As active safety technologies to assist in collision avoidance, chassis control technologies such as ESC is included, but in this section, collision damage mitigation braking will be introduced as a representative system that predicts and judges based on information from external sensors and controls the vehicle.

The collision damage mitigation braking is a system that controls the brakes to execute an emergency braking when a collision is imminent due to a driver's forward inattention, etc., assisting in collision avoidance and damage mitigation. It was first put into practical use in the world in 2003. The target collision scenarios have been gradually expanded with technological advancements, but the initial target was rear-end collisions with preceding vehicles, and it was called the rear-end collision mitigation braking system. Figure 8.5 shows the operation of the system. If a driver does not brake even when approaching a preceding vehicle due to some reasons such as forward inattention, the system first alerts the driver with a display and a buzzer sound as a primary warning, and then softly pulls in the seatbelt with an electric pretensioner while applying a soft brake as a secondary warning with tactile feedback. If the driver's operation is still not in time, the system assists in collision avoidance and damage mitigation by strongly pulling in the seatbelt and applying a strong brake.

As for the concept of assistance, the operation timing is set so that no warning occurs during normal driving, with the driver's operation at the center. Furthermore, for strong brake operation, it is set at a timing close to the physical limit where the expected collision can be avoided with maximum vehicle motion performance.

A millimeter-wave radar was adopted as an external sensor to detect preceding vehicles. In recent years, recognition technology using camera images has evolved, and the target objects have been expanded to pedestrians, bicycles, motorcycles, etc., which has expanded the types of collision scenarios covered and greatly improved the traffic collision reduction effect.

In Japan, the collision damage mitigation braking has been mandatory since 2021, and it has become the most representative collision avoidance support function worldwide.

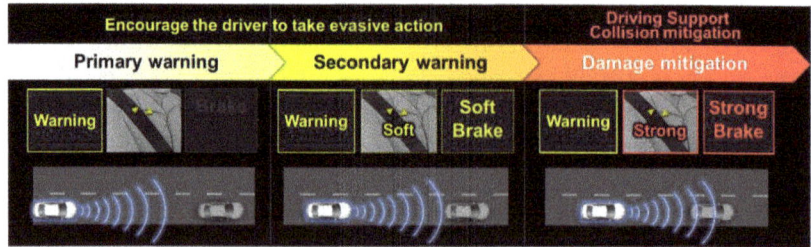

Fig. 8.5 Operation of collision mitigation braking system

8.4 Driver Assistance

The driver assistance function corresponds to level 1 or level 2. Various functions have already been put into practical use and are still being expanded, but this section explains ACC and LKA as representative functions.

ACC is called a constant speed driving/distance control system in the definition of ASV (Advanced Safety Vehicle) technology by the Ministry of Land, Infrastructure, Transport and Tourism [11]. Traditional cruise control controls driving force to maintain the set vehicle speed for constant speed driving. In addition to this function, ACC also detects a preceding vehicle with a millimeter-wave radar or a camera, and if the preceding vehicle's speed is lower than the set speed, it slows down the vehicle and controls it to follow while maintaining the predetermined distance from the preceding vehicle [12]. In most traffic situations on expressways, there is no need for accelerator and brake operations, greatly reducing driving burden on the driver.

LKA, known as lane keeping assist control system in ASV technology definition [11], recognizes the lanes in front of the vehicle through a camera and assists in steering operation to maintain the center of the lane, aiming to reduce driving burden on the driver [13]. In many commercially available models, the driver needs to keep their hands on the steering wheel, and if they let go, a warning is activated by the HMI. In some recent models, by utilizing high-precision maps, they have achieved assistance that can maintain the lane even if the hands are let go. However, even such systems are level 2 driver assistance, and the driver always needs to monitor the surroundings for safety. A camera is installed inside the car to monitor the driver's condition, and if the driver is not properly monitoring the surroundings, the system alerts the driver.

8.5 Automated Driving for Personal Cars (Level 3)

As for automated driving systems for personal cars, in November 2020, the world's first level 3 automated driving system was approved in Japan [14], and sales began in March 2021. This section explains the automated driving system installed in this mid-size passenger car.

8.5.1 System Configuration

Figure 8.6 shows the configuration of the automated driving (level 3) system.

The system is equipped with a global navigation satellite system (GNSS) to detect the vehicle's position. Although GPS is famous as a GNSS, it receives signals from multiple satellite systems, including Japan's Quasi-Zenith Satellite System

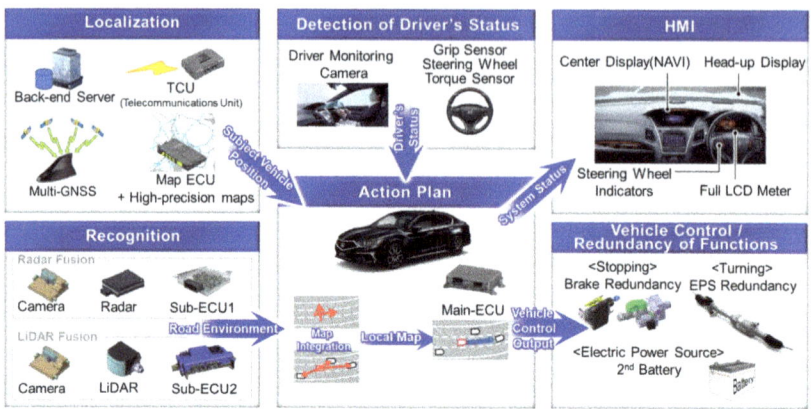

Fig. 8.6 Configuration of the automated driving (level 3) system

"Michibiki" (QZSS), to accurately detect the vehicle's latitude and longitude information. The map ECU stores high-precision maps of all motorways in Japan and identifies the vehicle's position on the map from the GNSS information. The high-precision map is regularly updated and transmitted to the vehicle via communication from the backend server.

For the function to recognize the outside around the vehicle, it is equipped with two front cameras, five millimeter-wave radars, and five LiDARs that sense all around. By placing two sub-ECUs (Electronic Control Units) to process this information, a redundant system is configured. It accurately recognizes road environment information such as the lateral position and shape of the lane in front of the vehicle, relative positions and relative speeds of surrounding vehicles.

As the function to detect the driver's status, it is equipped with a driver monitoring camera to monitor the driver's attention, a grip sensor to detect the steering operation state, and a steering wheel torque sensor.

Based on information about the vehicle's position, external recognition, and driver status detection, the main ECU constructs a local map of the vehicle's surroundings, creates the optimal action plan according to traffic conditions, and outputs control instructions to the vehicle control means with the calculated trajectory and the vehicle speed profile as the target.

The system's operating status is clearly communicated to the driver through visual, auditory, and tactile HMIs. In particular, LED indicators are installed on the steering wheel to intuitively convey the need for steering operation.

Regarding vehicle control, redundancy is aimed for in turning function of EPS (Electric Power Steering) and stopping function of brakes. Also as for power supply, a secondary battery is installed for duplication.

8.5.2 Provided Functions

The system provides advanced driver assistance functions in addition to the automated driving function. The main functions provided. When the driver enters a motorway and activates the system, the previously mentioned ACC and LKA functions are provided first. If the predetermined conditions are met, such as accurate identification of the vehicle's position, the lane-keeping assist function with hands-off function is activated, allowing the driver to take their hands off the steering wheel, greatly reducing the driving burden. Furthermore, if the advanced lane change assist function with hands-off function is activated, if there is a slower vehicle ahead and no approaching vehicles are detected from the rear, the system prompts the driver to check safety while controlling the steering to execute a lane change. After overtaking the slower vehicle, a lane change back to the original lane is performed. Also during this process, it is possible for the driver to keep their hands off the steering wheel. Although these are advanced functions, they are still level 2 driver assistance, so the driver needs to constantly monitor the surrounding traffic environment.

When a traffic condition become congested and the specified condition of the Operational Design Domain (ODD) are met, the Traffic Jam Pilot function is activated. This function is level 3 automated driving, and the system performs the Dynamic Driving Task (DDT). The driver no longer needs to monitor the surrounding traffic environment, and it is possible to watch videos such as TV or DVD on the center display. When the traffic jam is resolved and the vehicle speed increases, it will deviate from the specified ODD, so the system communicates the Request to Intervene (RTI) to the driver through the HMI. The driver needs to take over the driving promptly.

8.5.3 Safety Assurance

The biggest challenge in realizing automated driving is assuring safety. Up to level 2, the driver was the authority of Dynamic Driving Task (DDT), but at level 3 and above, all DDTs within the Operational Design Domain (ODD) are performed by the automated driving system. Regarding the safety required of the system, the question of "How safe is safe enough?" is still an ongoing global discussion. In Japan, the "Safety Technology Guidelines for Automated Vehicles" [15] issued by the Ministry of Land, Infrastructure, Transport and Tourism in 2018 stated that " The automated driving system shall not cause any personal injury accidents that are reasonably foreseeable and preventable." as the fundamental concept regarding safety of the automated driving system. By applying a highly reliable external recognition system with cameras, millimeter-wave radars, LiDARs, a redundant design of the entire system, and thorough validation by completed vehicles and simulations, it has become possible to meet the requirements for the safety.

For level 3 automated driving, ensuring safety during a handover of driving from the system to the driver is also an important issue. In the deployed system, the risk minimization control has been developed, assuming situations where the driver may not be able to respond to the Request to Intervene (RTI). Visual and auditory RTIs are escalated with a haptic warning (seatbelt retraction). If there is still no operation by the driver, the system notifies the surroundings by flashing hazard lights and honking a horn, while gradually slowing down and stopping the vehicle. If a safe space is confirmed on a road shoulder, the system controls the steering to enter the shoulder and stop.

Regulations and standards related to automated driving technology will be explained in Sect. 8.7, but there is still no established concrete one for the development process of automated driving and the validation of safety. The development had been carried out with reference to the concept of existing safety standards, while anticipating upcoming safety-related regulations and standards under investigation.

The basic policy of safety argumentation is to prove two things: (1) that the system operates within the ODD and the residual risk is within an acceptable range and (2) that the system does not operate outside the ODD. By properly designing and limiting the ODD, it is possible to minimize the residual risk.

For the concept of "shall not cause any personal injury accidents that are reasonably foreseeable and preventable," the approach of SOTIF (ISO21448: Road vehicles—Safety of the intended functionality) was introduced in advance, and "unknown" areas were thoroughly transitioned to "known" areas, clarifying the "reasonably foreseeable" range. Furthermore, by transitioning a "known" and "hazardous" area to a "not hazardous" area, it was argued that "shall not cause any personal injury accidents that are preventable."

In the validation phase, it was conducted through simulations and completed vehicles. Simulations included driving simulators, Model in the Loop (MIL) simulations, and Hardware in the Loop simulations. It was executed by combining those simulation methods. Particularly in MIL, all motorways in Japan that are the target of the ODD were reconstructed as road environment models in a virtual space, and a vast number of verification patterns were generated by combining assumed traffic flow patterns and traffic flow parameters, and validation was carried out. For completed vehicles, based on the operation verification at the dedicated test course, about 1.3 million km of driving was performed nationwide as public road validation. The data obtained in the public road validation was fed back to the simulation, ensuring the coverage of the validation.

8.6 Automated Driving for Mobility Services

While the practical application of automated driving system level 3 has mainly developed as a personal car, automated driving system level 4, which does not assume the existence of a driver, has been studied with a focus on application to

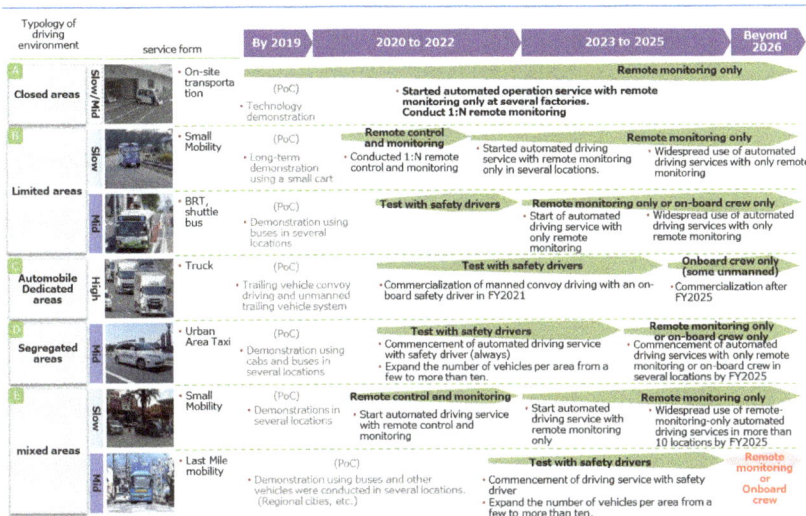

Fig. 8.7 Japanese government roadmap for realization and spread of driverless mobility services [16]

mobility services. Figure 8.7 shows the types of autonomous driving mobility services compiled by the Ministry of Economy, Trade and Industry [16].

As shown here, when trying to apply autonomous driving that does not assume the existence of a driver to mobility services, it is classified into seven types depending on the form of the service and its provision range.

In addition, the Japanese government has set a goal for the social implementation of mobility services using automated driving system level 4 to about 50 places nationwide by 2025, and about 100 places by 2027, and has shown a medium- to long-term social implementation plan and government promotion system that is based on the basic concept of "from point to line/area, from demonstration to implementation" in the "National Comprehensive Development Plan for Digital Lifelines."

8.6.1 Autonomous Driving Mobility Services in Rural Areas

A representative example of the social implementation of autonomous driving mobility services in rural areas is the "ZEN drive" in Eiheiji Town, Fukui Prefecture (Fig. 8.8) [17]. This project has been promoted jointly by the Ministry of Economy, Trade and Industry and the Ministry of Land, Infrastructure, Transport and Tourism since fiscal 2021 as Theme 1 of the "Research and Development and Social Implementation Project for Advanced Mobility Services such as Automated Driving System Level 4", and demonstration experiments for social implementation have

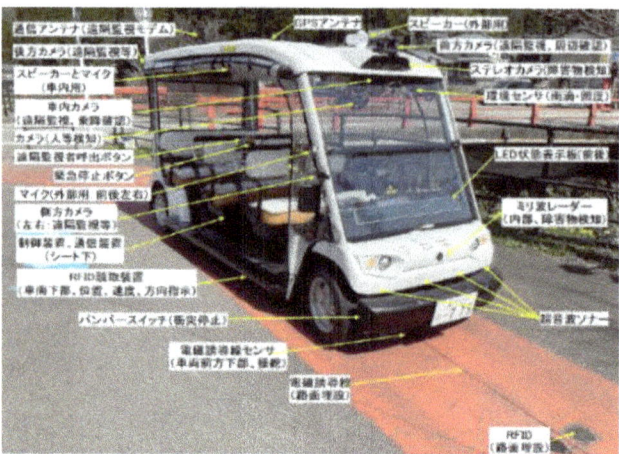

Fig. 8.8 The test vehicle given operational design domain [17]

been conducted. In March 2023, it obtained approval as an automated driving device under the Road Transport Vehicle Act, and in May 2023, it obtained permission for specific autonomous operation under the Road Traffic Act, leading to the start of the project. Inside the vehicle used for specific autonomous operation, no driver or other person rides, and three vehicles can run at the same time by remote monitoring. This example corresponds to the slow speed type of limited area B as among the aforementioned seven types.

8.6.2 Autonomous Driving Mobility Services in Urban Areas

As of the end of 2023, there are no examples of autonomous driving services in urban areas that have reached societal implementation. Instead, bus and taxi operators and emerging technology development companies are conducting demonstration experiments on public roads. A notable example is the demonstration experiment on public roads conducted by the "Project for the Construction of Services Utilizing Automated Driving Technology in the Waterfront Subcenter in Fiscal Year Reiwa 5" promoted by the Tokyo Metropolitan Government [18].

In addition to the above, demonstration experiments are being conducted in various regions and cities. An example of an activity being carried out with commercialization in plan is the case where Honda R&D Co., Ltd. announced in October 2023 that it has established a joint venture with General Motors and GM Cruise Holdings to aim for the launch of an autonomous taxi service in early 2026 [19]. This example corresponds to the middle speed type of mixed area E as among the aforementioned seven types, and it is expected that the societal implementation of autonomous driving services in urban areas will become more substantial due to this announcement.

8.7 Regulations and Standards Related to Automated Driving Technology

8.7.1 International Standardization Activities Related to Automated Driving

In the research and development and societal implementation of automated driving, it is important to establish socially recognized technical standards and to implement and experiment based on them. This section introduces representative international standardization activities.

The International Organization for Standardization has two technical committees, TC22, which deals with automobile standards, and TC204, which deals with Intelligent Transport Systems (ITS). TC22 mainly deals with design standards and test methods, and Japan led the establishment of ISO 34502, a safety verification method for automated driving in SC33 WG9 under it. This standard provides guidance on a scenario-based safety evaluation framework for Automated Driving Systems (ADS). The framework details the scenario-based safety evaluation process applied during product development. TC204 mainly deals with functional requirements of vehicle systems, and Japan also led the establishment of ISO 23792 Motorway Chauffer System (MCS) as an automated driving system level 3 in WG14 under it. The Motorway Chauffer System (MCS) is defined as a system that performs automated driving system level 3 on access-restricted highways, assuming the presence of a fallback-ready user (FRU), and the standard describes the system characteristics, system state/transition conditions, system functions, etc., of the MCS framework.

As can be seen from these international standardization activities for automated driving, Japan is also leading international discussions in a leading position.

8.7.2 Trends in Harmonization of International Regulation Related to Automated Driving

In 2015, the IWG was established with the aim of deregulating ACSF (Automatic Command Steering Function), and discussions began. Here, the steering functions were categorized from A to E and organized into lane keeping (handle holding is category B1, handle non-holding is category B2), lane changing (driver instruction is category C, system judgment is category D), and continuous automatic steering (category E). However, only categories B1 and C were adopted in UNECE R79 in November 2018, and no regulations were established for categories B2, D, and E, which have a high degree of automation.

In 2018, the GRRF under the UNECE's WP29 (World Forum for Harmonization of Vehicle Regulations) was reorganized into the GRVA, which deals with automated/autonomous and connected vehicles. In June 2019, a framework document for autonomous driving (international guidelines and regulation setting schedule for automated/autonomous vehicles) was agreed upon, accelerating discussions on automated driving. In particular, the international technical level for automated driving was documented by the FRAV, which establishes guidelines indicating the functional requirements of automated driving, and the VMAD, which discusses the safety evaluation technology of automated driving. This activity led to the issuance of R157, a regulation for "an automatic operation device with a function to keep the vehicle in the lane during operation on highways, etc." at the WP29 plenary meeting held in June 2020.

While harmonization of international regulation activities for automated driving are progressing, the threat of cybersecurity is expanding, and in response to the situation where OTA (Over The Air) technology, which transmits and receives data via general wireless communication such as smartphones, is becoming widespread for automobiles, WP29 has also started to establish regulations for automobiles. As a result, UNECE R155, which defines the cybersecurity and cybersecurity management system (CSMS) of automobiles, and UNECE R156, which defines the software update and software update management system (SUMS) of automobiles, came into effect in January 2021.

R157, R155, and R156 are major international regulations related to automated driving, and in Japan, the safety requirements of the Road Transport Vehicle Act are harmonized with them.

In parallel with the establishment of basic regulations for automated/autonomous vehicles by the UNECE's harmonization activities of international regulations, the European Commission adopted "New Rules for Improving Traffic Safety in the EU and Realizing Fully Automated Vehicles" in July 2022. This new European standard focuses on automated vehicles (automated driving system level 3) that replace drivers, especially on highways, and fully automated vehicles (automated driving system level 4) such as urban shuttles and robot taxis, and sets technical requirements such as test procedures, cybersecurity requirements, data recording rules, safety performance monitoring, and accident reporting requirements by fully automated vehicle manufacturers.

In response to such harmonization activities of international regulations and international standardization activities, the Japanese government has established the Automated Driving Regulation Research Institute within JASIC (Japan Automobile Standards Internationalization Research Center). The government and industry are working together on various issues related to international harmonization of automated driving regulation to help Japan stay competitive in the global market.

8.8 Initiatives of the Japanese Government on Automated Driving

In "Declaration to Be the World's Most Advanced IT Nation" [20] decided by the Cabinet in 2013, it was stated that trials of automated driving systems would start in the 2020s and that the world's safest road traffic society would be realized by 2020.

8 Automated Driving and Driver Assistance

Events such as former Prime Minister Abe test-driving vehicles equipped with automated driving technologies around the National Diet Building have also been a catalyst, and the Japanese government has strengthened its efforts toward automated driving technology. In 2014, "Automated Driving System" was started as one of the themes of the Cross-ministerial Strategic Innovation Promotion Program (SIP), which is promoted by industry-academia-government collaboration. In the same year, the IT Strategy Headquarters, headed by the Prime Minister, formulated the "Government-Private ITS Concept/Roadmap," and the schedule for the early deployment of automated driving was shown. In 2015, the Ministry of Economy, Trade and Industry, in collaboration with the Ministry of Land, Infrastructure, Transport and Tourism's Automobile Bureau, launched the "Automated Driving Business Study Group" [21], and discussions on analysis of foreign situations, strategic division of competitive and cooperative areas, etc., are being conducted among various stakeholders.

8.8.1 Legal System Development for Practical Application of Automated Driving

In the revised "Government-Private ITS Concept/Roadmap 2016," it was indicated that the goal is to achieve level 3 on highways for personal cars around 2020 and to realize driverless automated driving mobility services (Level 4) in limited areas and dedicated spaces by 2020. In line with this goal, the "Outline of Legal System Development for Automated Driving" [22] was issued in 2018, and the legal system development necessary for the social implementation of automated driving was accelerated.

Toward the realization of level 3 personal cars, two amendments to the Road Traffic Law for drivers and the Road Transport Vehicle Law for vehicles were enacted in April 2020, making it the first time in the world that legal system preparations were completed. As a result, in November 2020, the world's first type approval of level 3 automated driving vehicles equipped with traffic jam pilot was given [23].

In addition, amendments to the law were made for driverless automated driving services, which were enacted in April 2023. In March 2023, vehicles to be operated as a driverless automated driving service in Eiheiji, Fukui Prefecture, were approved as automated driving vehicles (Level 4), and the service started in May 2023.

8.8.2 Initiatives of the Strategic Innovation Promotion Program (SIP)

SIP is a national project that promotes research and development in a comprehensive manner, looking toward practical application and commercialization, while promoting industry-academia-government collaboration beyond the framework of

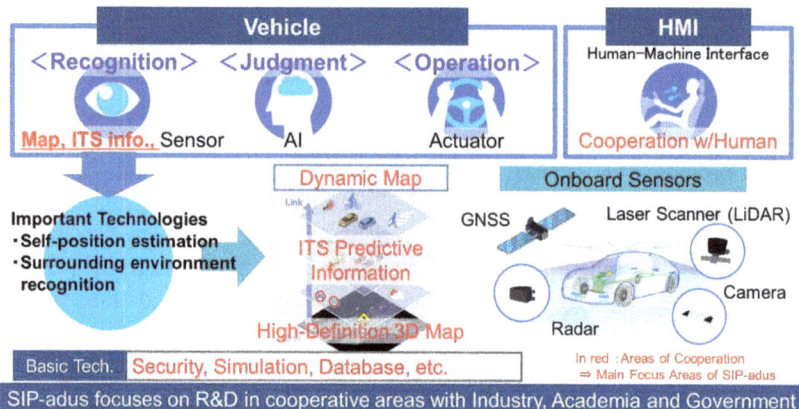

Fig. 8.9 Scope of the first phase of SIP automated driving system [25]

conventional ministries and agencies, with the Cabinet Office's Council for Science, Technology and Innovation as the command tower. Regarding automated driving, efforts were made over a period of 9 years from fiscal 2014 to fiscal 2022 through the first and the second phases of the SIP [24].

First, in the first phase, competitive and cooperative areas related to automated driving technologies were organized, and the SIP's efforts were focused on the cooperative areas shown in Fig. 8.9 [25].

Regarding the dynamic map based on high-precision digital maps, Dynamic Map Platform Planning Co., Ltd. was established as the outcome of the SIP's efforts and later commercialized as Dynamic Map Platform Co., Ltd. High-precision 3D maps were developed for all motorways in Japan, covering about 30,000 km and were adopted for a level 2 advanced driver assistance system in 2019 and for a level 3 automated driving system in 2021.

In the second phase, field operational tests, technological development, fostering societal acceptance, and international cooperation were promoted as the four pillars. Among these, the technological development included traffic environment information composed of signal information, V2X information, vehicle probe information, etc., safety verification environment in virtual space, HMI, security, etc. The purpose of the safety verification environment in virtual space was to reconstruct various driving environments in virtual space and verify safety through simulation [26]. Especially, to improve reconstruction accuracy of sensor output information, sensor models and environment models were constructed based on each sensor's detection principle and detailed measurements of reflection characteristics of light and radio waves, and consistency of actual sensor outputs and model sensor outputs was verified [27]. This allows for the simulation and assessment of unwanted reflection output from a millimeter-wave radar in tunnels, which is called the ghost phenomenon, and the outputs of sensors that use light under various difficult environmental conditions, such as rain, wet roads, backlight, and so on. The DIVP (Driving Intelligence Validation Platform) consortium supported this advancement and then turned it into a business as V-Drive Technologies Co., Ltd.

8.9 Future Evolution

For personal cars, it is expected that automated driving systems will evolve from the level 3 automated driving function during traffic congestion on motorways to level 3 automated driving functions throughout motorways. While automated driving systems are likely to take a while to spread, the technologies created along the way are used for fast improvement of driver assistance systems. The collision avoidance function is expected to further reduce traffic collisions as collision scenarios in which it can effectively function are expanded. The driver assistance function is expected to evolve from support on expressways to seamless support on all kinds of roads, contributing to reducing driving burden for drivers, especially elderly drivers and novice drivers.

For automated driving for mobility services, it is highly expected as a solution to the serious driver shortage problem in public transportation and logistics sectors. However, level 4 automated driving with complete unmanned operation needs a large investment, requiring highly advanced technologies and large-scale operation systems. Expanding the business in urban areas where there is a lot of demand may be an option, but in rural areas where public transportation is hard to sustain, a form of operation that requires less investment and operational costs is needed. Together with operations that involve auxiliary support staff rather than being completely unmanned, securing driving space with the understanding of residents, and cooperation with other businesses, it is necessary to introduce mobility services that suit the needs of each area.

To accelerate the adoption of automated driving technologies in society, it is crucial to strengthen the cooperation among industry, academia, government, and the public, covering aspects such as the cultivation of societal acceptance, legal frameworks including ethical considerations, traffic system design, and sustainable business models. We look forward to realizing a society where everyone can travel anywhere, anytime, with safety and peace of mind, without traffic collisions, by advancing and expanding the use of automated driving and driver assistance technologies.

References

1. Weber M Where to? A History of Autonomous Vehicles - CHM (computerhistory.org). Accessed 31 Oct 2023
2. Report on the First International Technical Conference on Experimental Safety Vehicles, National Highway Traffic Safety Administration (U.S. Department of Transportation), 1971
3. Development History of AHS and ASV, Road Bureau, Ministry of Land, Infrastructure, Transport and Tourism, Development History of AHS (Advanced Cruise-Assist Highway Systems) (mlit.go.jp). Accessed 31 Oct 2023
4. Furukawa O et al (1996) Overview of Honda AHS. Honda R&D Tech Rev 8:12–20
5. The DARPA Grand Challenge: Ten Years Later, Defense Advanced Research Projects Agency. Accessed 31 Oct 2023

6. DARPA Urban Challenge, Defense Advanced Research Projects Agency. Accessed 31 Oct 2023
7. Taxonomy and Definitions for Terms Related to Driving Automation Systems for On-Road Motor Vehicles J3016_202104, SAE International, 2021
8. Taxonomy and Definitions for Terms Related to Driving Automation Systems for On-Road Motor Vehicles (SAE J3016:2021 Japanese Reference Translation), JASO Technical Paper No. TP-18004-22, The Society of Automotive Engineers of Japan, 2022
9. Government-Private ITS Concept/Roadmap 2017, Prime Minister's Office, 2017
10. Japan's First! Level 4 Automated Driving Mobility Service Has Started, Ministry of Economy, Trade and Industry, Japan's First! Level 4 Automated Driving Mobility Service Has Started (METI/Ministry of Economy, Trade and Industry). Accessed 31 Oct 2023
11. Overview of Major ASV Technologies (from the 6th Report Document Edition by the 6th Advanced Safety Vehicle (ASV) Promotion Plan), Advanced Safety Vehicle Promotion Study Committee, Ministry of Land, Infrastructure, Transport and Tourism, 2021
12. Adaptive Cruise Control (ACC), Honda Motor Co., Ltd., Adaptive Cruise Control (ACC) | Technology | Official Honda Site (global.honda). Accessed 31 Oct 2023
13. Lane Keeping Assist System (LKAS), Honda Motor Co., Ltd., Lane Keeping Assist System (LKAS) | Technology | Official Honda Site (global.honda). Accessed 31 Oct 2023
14. World's first! We have given the type approval of automated driving vehicle (Level 3), Road Transport Bureau of the Ministry of Land, Infrastructure, Transport and Tourism, Press Release: World's first! We have given the type approval of automated driving vehicle (Level 3) - Ministry of Land, Infrastructure, Transport and Tourism (mlit.go.jp). Accessed 31 Oct 2023
15. Safety technology guidelines for automated vehicles, Road Transport Bureau of the Ministry of Land, Infrastructure, Transport and Tourism, 2018
16. Automated Driving Business Study Group (2020) Report and policy for realizing automated driving. Version4.0 Report Summary, Ministry of Economy, Trade and Industry
17. Obtained the first domestic approval for Level 4 automated vehicles with remote monitoring only, National Institute of Advanced Industrial Science and Technology, AIST: Obtained the first domestic approval for Level 4 automated vehicles with remote monitoring only (aist.go.jp). Accessed 31 Oct 2023
18. Project on the Construction of Services Utilizing Automated Driving Technology in the Rinkai Subcenter in FY2023, Nippon Koei Co., Ltd., Project on the Construction of Services Utilizing Automated Driving Technology in the Rinkai Subcenter (autonomouscar-tokyo.jp). Accessed 31 Oct 2023
19. Honda, GM and Cruise Plan to Begin Driverless Ridehail Service in early 2026, Honda Motor Co., Ltd., Honda Global Corporate Website. Accessed 31 Oct 2023
20. Declaration to Be the World's Most Advanced IT Nation, Strategic Headquarters for the Promotion of an Advanced Information and Telecommunications Network Society, 2013
21. Automated Driving Business Study Group (n.d.) Ministry of Economy, Trade and Industry, Automated Driving Business Study Group (METI/Ministry of Economy, Trade and Industry). Accessed 31 Oct 2023
22. Outline of legal system development for automated driving, Prime Minister's Office, 2018
23. Japan's first! About the approval of automated vehicles (Level 4) that do not require a driver, Road Transport Bureau of the Ministry of Land, Infrastructure, Transport and Tourism, Press Release Material: The first in Japan! About the approval of automated vehicles (Level 4) that do not require a driver - Ministry of Land, Infrastructure, Transport and Tourism (mlit.go.jp). Accessed on 31 Oct 2023
24. Automated Driving (System and Service Expansion) SIP-adus, Cabinet Office, SIP-adus Automated Driving for Universal Service. Accessed on 31 Oct 2023
25. Kuzumaki S (2017) Automated driving system. SIP Symposium 2017, Cabinet Office

26. Hosaka O et al. (2022) Technological Development and Education for Enhanced Safety (Overview), SIP 2nd Phase: Automated Driving for Universal Services Final Results Report (2018–2022), pp104–107, Cabinet Office, – SIP-adus Automated Driving for Universal Service. Accessed 31 Oct 2023
27. Inoue H (2022) Development of Driving Intelligence Validation Platform (DIVP®) for Automated Driving Safety Assurance, SIP 2nd Phase: Automated Driving for Universal Services Final Results Report (2018–2022), pp 108–120, Cabinet Office, 2022. – SIP-adus Automated Driving for Universal Service. Accessed 31 Oct 2023

Open Access This chapter is licensed under the terms of the Creative Commons Attribution-NonCommercial-NoDerivatives 4.0 International License (http://creativecommons.org/licenses/by-nc-nd/4.0/), which permits any noncommercial use, sharing, distribution and reproduction in any medium or format, as long as you give appropriate credit to the original author(s) and the source, provide a link to the Creative Commons license and indicate if you modified the licensed material. You do not have permission under this license to share adapted material derived from this chapter or parts of it.

The images or other third party material in this chapter are included in the chapter's Creative Commons license, unless indicated otherwise in a credit line to the material. If material is not included in the chapter's Creative Commons license and your intended use is not permitted by statutory regulation or exceeds the permitted use, you will need to obtain permission directly from the copyright holder.

Chapter 9
Traffic Psychology

Kazumitsu Shinohara and Kazuko Okamura

9.1 Psychological Processes of Traffic Participants

When humans, as traffic participants, navigate the road, they gather information from the external environment to understand the situation, decide how to act according to their intentions, and operate the steering wheel, pedals, etc., if they are driving a car or riding a bicycle. This series of psychological processes is carried out at various levels of consciousness, and humans cannot consciously monitor all of them. This process is schematically represented in Fig. 9.1. In Japan, those seeking to obtain a driver's license at a driving school learn at the outset that driving a car is a repetitive process of cognition, judgment, and operation. This process of behavior is not limited to driving but is common to all human behavior of moving around.

9.1.1 Model of Traffic Behavior

When considering behavior in actual traffic situations, it is necessary to consider not only the simplest division into cognition, decision-making, and operation but also the human factors that influence this process. Moreover, this behavior includes multiple levels of behavior. Figure 9.2 shows a behavior control model [1] that adds the roles of attention and cognitive control to the Skill-Rule-Knowledge (SRK) model [2], which is often used as a model of human behavior in human factors research. This model shows that human behavior is controlled at three levels. In skill-based behavior, reactions mostly occur automatically in response to information obtained

---

K. Shinohara (✉)
Graduate School of Human Sciences, The University of Osaka, Suita, Japan
e-mail: kaz.shinohara.hus@osaka-u.ac.jp

K. Okamura
Department of Traffic Science, National Research Institute of Police Science, Chiba, Japan

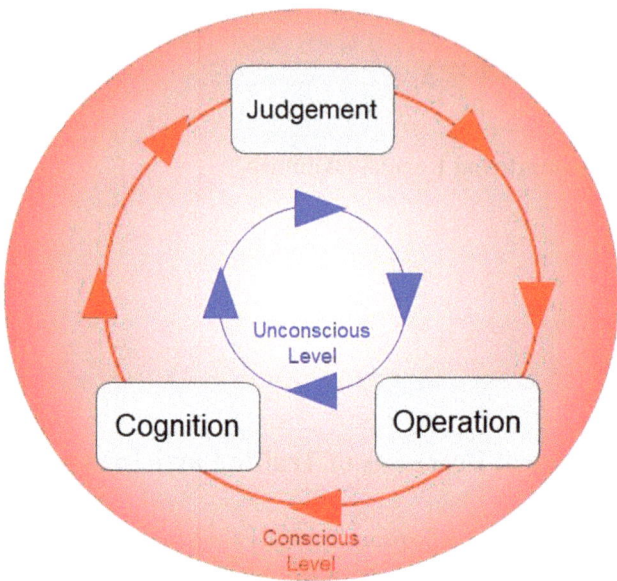

Fig. 9.1 Psychological process of driving behavior

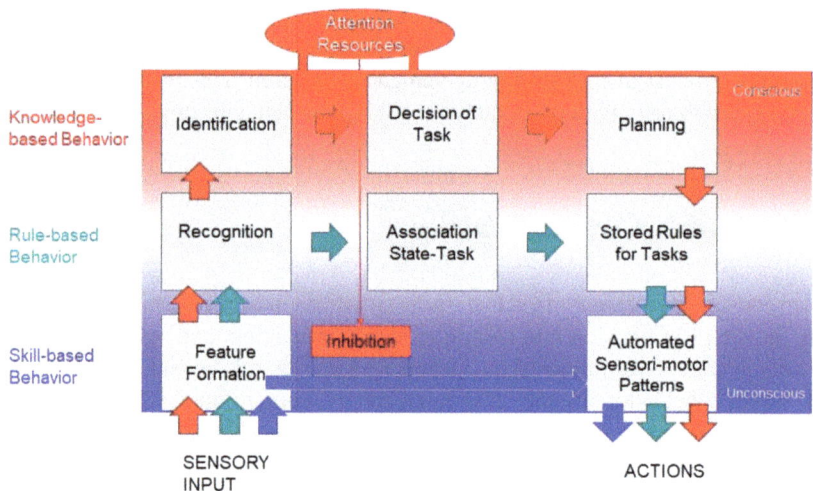

Fig. 9.2 Behavior control model based on Rasmussen's SRK model

from the external environment. For example, reflexively braking when a pedestrian steps into the lane ahead or when the preceding vehicle suddenly brakes would fall into this category. Rule-based behavior requires selecting actions based on the content of the information, where rules for how to respond and recalling the content of the response take place. For instance, stopping at a red traffic light is an action based on rules. Such signal-based actions are simple and familiar, and thus can be performed almost unconsciously; however, when faced with an unfamiliar sign, some thought may be required on how to act. Therefore, the level of consciousness in rule-based behavior varies. In knowledge-based behavior, it is necessary to interpret the content of information, think about what an appropriate decision would be, and consider what actions should be taken to realize that decision. For example, deciding which direction to go based on guide signs while driving in an unfamiliar area and navigating through intersections for the first time would fall into this category. It can be said that driving behavior consists of these multiple levels of actions combined. Additionally, according to dual-process theory [3], which posits two methods of thinking, unconscious and conscious processes, the former is called System 1 and the latter System 2. System 1 is characterized as intuitive, effortless, fast, context-dependent, and rigid, while System 2 is logical, requires attention and effort, slow, and flexible [4]. The cycle of cognition, decision-making, and operation occurs with various types of behaviors shown in these behavioral models, and various levels of consciousness control these actions. The cycle of cognition, decision-making, and operation includes behaviors at each level indicated in these behavior models, and the levels of control for these behaviors are also diverse.

In driving behavior, the driver's attention plays a crucial role in safety. However, while attention directly influences actions when they are conscious, its impact is minimal when actions are performed unconsciously. Considering that many human behaviors are controlled at an unconscious level, it is not enough to be able to drive safely consciously; safe driving must also be possible under unconscious behavioral control. Generally, as skills mature through repeated experience and training, behavior gradually becomes automatic and unconscious. Drivers automate their operational driving skills through training and accumulated driving experience. In addition, drivers need to acquire and automate not only driving operational skills, but also psychological skills related to safety, such as hazard prediction. Furthermore, road environment maintenance should aim to reduce the need for drivers to consciously pay attention, naturally drawing drivers' attention and inducing behavior.

Moreover, when capturing driving behavior more comprehensively, it can be understood not merely as a simple cycle of cognition, decision-making, and operation but as directed by various levels of decision-making. This hierarchical concept of driving behavior is indicated within the goals and content of driver education (GDE framework) [5]. This model defines layers of driving behavior as: (1) operation of the vehicle, (2) understanding of traffic situations, (3) driving goals and context, (4) personal goals in life and living skills, and (5) social environment. The cycle of cognition, decision-making, and operation (Fig. 9.1) corresponds to the first and second layers of this model; however, as driver assistance and autonomous driving technology are introduced, these layers may be taken over by systems, while human drivers will be responsible for the third to fifth layers.

9.1.2 Attention and Cognition in Traffic Behavior

Many traffic accidents are caused by human factors, and a more detailed investigation of the causes reveals that inappropriate attention and errors in judgment lead to accidents. When humans act on the road, they make decisions about what actions to take while obtaining the necessary information from the surrounding environment and moving their bodies. In this process, the selection of information and the maintenance of cognitive information processing for action are necessary, and it is essential to use attention and cognitive functions appropriately.

Attention can be considered from two perspectives: its function as a filter that selects information input to the cognitive system (selective attention) and its function as a mental resource necessary for processing the selected information (attentional resources) (Fig. 9.4) [7]. First, regarding the function of selection, the human cognitive system has limited processing capacity, so it is necessary to select the information currently needed. Most of the information required for driving a car is provided visually. When driving, it is necessary to obtain information from a wide range and selectively obtain detailed information through central vision with high visual acuity by directing the gaze to particularly important areas. This selection of information is influenced by anticipation and value judgments. To visually perceive information, it is insufficient to simply direct the gaze; without the intention of detecting something, the phenomenon of inattentional blindness [8] may occur, where an object that should be visible is overlooked, and accidents may occur due to "Looked But Failed to See" [9] (Fig. 9.3).

Next, the selected information is kept in an activated state in working memory, but this requires an appropriate allocation of attentional resources. Attentional

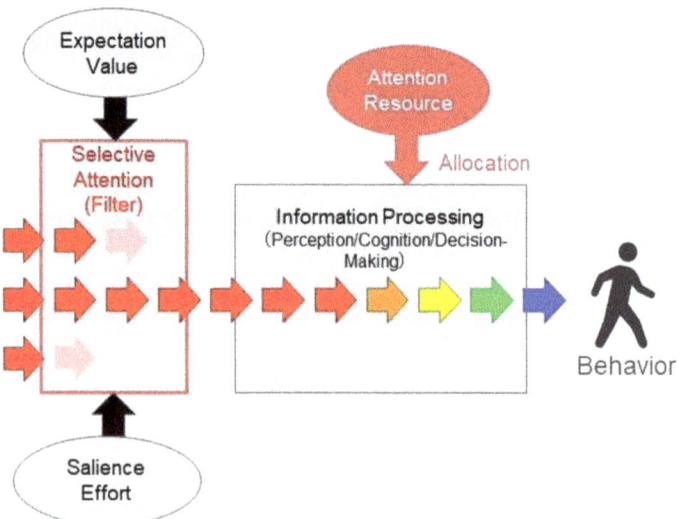

Fig. 9.3 The role of attention in human information processing

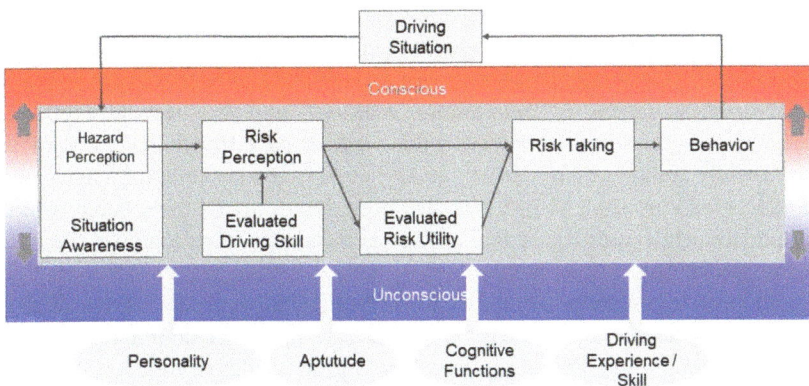

Fig. 9.4 Driving hazard perception, risk perception, and risk-taking. (Modified based on Renge [6])

resources have a capacity limit, and their allocation is determined by the intention of the behavior at any given time, the difficulty and importance of the task, and other factors. In so-called absent-minded driving, where the driver is thinking about something else or performing a task unrelated to driving while their gaze is directed forward, they may overlook dangers ahead or make driving errors. This can be interpreted because of insufficient attentional allocation to driving, leading to a decline in the information processing processes related to driving.

While driving, drivers move their gaze to look ahead and around their path, but detailed visual information can be obtained within the useful field of view. The useful field of view refers to the range around the fixation point where information processing is possible, and its size varies from about 4 to 20 degrees depending on various psychological and environmental factors. Miura [10] showed that drivers' useful field of view changes in response to driving conditions such as road congestion and that this change controls how attentional resources are allocated to the fixation target according to the situation. Additionally, the size of the useful field of view narrows with age, and a correlation has been found between the size of the useful field of view in older adults and their accident experience. A driving ability test based on the measurement of the useful field of view (UFOV test) has been developed [11, 12].

Distracted driving, where drivers' attention is diverted to objects or tasks other than driving, resulting in insufficient attention to driving, has become a significant concern. According to the National Highway Traffic Safety Administration (NHTSA), in 2021, distracted driving was associated with 8% of fatal crashes and 13% of reported motor vehicle crashes in the United States, resulting in 3522 deaths and 362,415 injuries [13]. Distracted driving has gained attention due to the proliferation of mobile devices, such as cell phones, leading to concerns about drivers engaging in phone conversations or sending text messages while driving. However, driver distraction is not limited to the use of mobile devices while driving; it can also occur when drivers engage in various other behaviors while behind the wheel.

A study [14] that recorded and classified various behaviors that drivers routinely engage in while driving using in-vehicle devices reported that driving behavior is affected while performing these behaviors. In recent years, walking while using mobile devices, known as "smartphone walking," has also become a problem. Studies on the use of mobile devices while walking have shown that using mobile devices while walking delays reactions to events in the surrounding environment and affects walking itself [15]. It has also been demonstrated that people who use smartphones while walking in real traffic environments have a higher risk when crossing roads [16].

In addition to attention, the ability to control cognitive processes that form the basis of traffic behavior also plays a crucial role in the psychological functions required while driving. This ability is referred to as executive function. Executive function includes cognitive flexibility to achieve goals, selection of necessary information and responses, and divergent thinking and fluency to identify strategies and various elements [17]. According to a representative model of executive function [18], it consists of shifting the contents of behavior, updating information in memory, and inhibiting automatic responses. Research has been conducted on the relationship between executive function and driving. For example, Adrian et al. [19] examined the correspondence between performance on various test tasks corresponding to executive function and driving behavior evaluations, showing that inhibitory function is related to driving behavior.

The increase in traffic accident fatalities among the elderly has become a social problem. In addition to the decline in physical abilities, the decline in executive function with aging is thought to be a contributing factor. It has been reported that there is a relationship between the decline in executive function and accident experience among elderly drivers aged 65 and older [20]. It has also been reported that age-related changes in cognitive function in the elderly are based on a general slowing of mental processing speed [21] and that the elderly have more difficulty coping with multitasking compared to younger individuals [22, 23]. Because driving behavior is inherently multitasking and requires appropriate speed and timing, it can be said that driving behavior becomes increasingly difficult for elderly drivers as their cognitive functions decline with age. However, elderly drivers can ensure safety by recognizing age-related changes in their physical and cognitive functions and taking compensatory actions. In fact, elderly drivers are engaging in various compensatory behaviors [24]. Providing technical support, diagnosis, and education for elderly drivers is important in terms of supporting the quality of life of the elderly from the perspective of mobility by ensuring their safety.

9.1.3 Hazard Perception

Hazard refers to a potential source of danger that could lead to accidents, and hazard perception is the act of detecting this hazard and recognizing its characteristics while driving. Hazards can be classified into three types: overt hazards that are visible and require a response, behavioral prediction hazards that are not currently visible but could become dangerous depending on future movements, and latent hazards that are locations where dangerous objects may be hidden from view [25]. While driving, it is necessary to detect and respond to these hazards at the appropriate timing. Hazard perception can also be considered as part of situation awareness, which is essential for ensuring safety while driving. Situation awareness refers to the perception of elements in the environment within a certain temporal and spatial range, the comprehension of these elements, and the projection of their status in the near future [26, 27].

As mentioned in the previous section, to appropriately select visual information and perceive hazards, drivers must understand what to look for in a given situation and be able to anticipate hazards. Hazard perception is based on acquiring sufficient knowledge about hazards through driving experience and education and being able to recall and utilize that knowledge while driving.

The ability to perform appropriate hazard perception is crucial for accident prevention. To assess hazard perception ability, various hazard perception tests have been developed and incorporated into licensing examinations in Western countries and Japan. A typical hazard perception test involves observing photographs, illustrations, video footage, or computer-generated imagery of the forward view from a driver's perspective and responding by pressing a button or other means when a hazard is detected. Hazard perception ability is evaluated based on detection response time and detection rate. In many cases, experienced drivers exhibit shorter detection response times and detect more hazards compared to novice drivers. Furthermore, research that recorded eye movements during hazard perception tests has shown that the difference in response speed between experienced and novice drivers is not merely due to perceptual detection ability, but rather due to differences in cognitive ability related to judging the level of danger posed by the detected hazards. A review study on hazard perception [28] indicated that it is possible to discriminate between experienced and novice, accident-prone and non-accident-prone drivers by their performance on hazard perception tests.

In addition to the assessment of hazard perception, hazard prediction education and training are also conducted to enhance the ability to perceive hazards. Specifically, there are methods that use two consecutive scenes illustrated to evaluate and educate hazard perception [29], those that present video footage of traffic environments and have participants identify hazards [30], and those that use tablet devices to present traffic situations and have participants touch the areas that require attention [31]. While all these methods share the common point of having participants discover hazards from driving scenes and using that as a basis for education, improvements continue to be made to create smarter methods in line with the progress of information and communication technology. Hazard perception involves recognizing situations based on knowledge about dangers, and for hazard perception to be performed appropriately, it is necessary to possess sufficient knowledge and for that knowledge to be appropriately recalled according to the situation. Moreover, to cope with complex traffic situations, it is necessary to know how the presence of hazards can lead to accidents, rather than simply training simple responses to individual hazards. For the driver, simply being able to detect a hazard is not enough; it is important to know useful cues for hazard detection and to understand the relationship between hazards and accidents.

Furthermore, with the increase in elderly drivers, research on age-related changes in hazard perception ability is also progressing. While sensitivity to hazard and risk increases with driving experience [30], reaction time in hazard perception lengthens with age [32]. There are still points that need to be examined, such as whether the changes in hazard perception test performance of elderly drivers are due to a decline in visual function itself or an overall decline in attention and cognitive function, whether they are influenced by the knowledge and cognitive functions that form the basis of hazard perception, or whether they are due to changes in judgment criteria rather than the hazard perception ability itself.

9.2 Risk Perception and Risk Taking

Not only driving behavior but also the act of "moving" on the road always involves the risk of accidents, and the degree of risk varies depending on the situation. Taking action despite the presence of risk can be regarded as risk-taking. Whether to engage in behavior that involves risk (risk-taking) or not (avoidance) is part of the judgment stage in human behavior control. When making this judgment, the following issues are considered: the degree of danger in the current situation (hazard perception), one's ability to act safely in that situation (self-assessment of driving ability), and the level of risk one is willing to accept and act upon (setting judgment criteria).

This process is illustrated in Fig. 9.4. When a driving situation is perceived, the hazards contained within it are perceived, followed by an assessment of the degree of danger those hazards pose in terms of leading to accidents (risk perception). Risk perception is influenced by the assessment of the inherent danger of the hazard itself and the assessment of one's own ability to cope with that hazard based on one's driving ability. Regarding self-assessment of driving ability (metacognition), there is a tendency to evaluate one's own driving skills higher than average, and self-assessment increases with driving experience. However, the change is rapid after obtaining a license, and the increase becomes gradual thereafter [33]. Moreover, older drivers maintain a high self-assessment despite an objective decline in their driving ability [24]. When one's skills are perceived as sufficient to handle the hazard, the risk is assessed as low.

Even when a risk is perceived, there may be instances where action is taken deliberately, but behind this lies the risk utility, which is the benefit gained from engaging in risky behavior. Risk utilities include stress relief, aggression, expression of independence, increased arousal level, urgency, defiance of authority, and praise from peers [6]. A human characteristic related to risk-taking is sensation seeking, which is the tendency to seek novel and changing stimuli [34]. A review [35] on the relationship between sensation seeking and driving behavior reports that a consistent positive correlation is found between sensation seeking and high-risk driving behaviors.

The risk homeostasis theory [36, 37] is a well-known theory regarding the decision to take or avoid risks. According to this theory, people have a "target risk level" and adjust their behavior to maintain that level of risk. Therefore, it is predicted that even if road environment improvements or driver skill training measures are implemented, if the target risk level remains the same, people will engage in more dangerous behaviors to adjust their actions to match the target risk level. Although this homeostasis process-based behavioral adjustment occurs at the individual level, this model is intended to explain the accident rate of all road users in each area, rather than individual behavior [36]. While this theory is widely known, its validity has also been debated.

Based on the judgment of whether to take risks or not, driving behavior is executed, and the results are reflected in the driving situation, leading to the next driving situation recognition. Personality, driving attitudes, cognitive functions, driving experience, and driving skills also influence each stage of this process. Furthermore, this series of processes is thought to be executed at various levels of consciousness, as mentioned earlier, which should also influence hazard perception, risk perception, and decision-making.

9.3 Interventions for Traffic Participants

This section explains theories and models related to changes in the attitudes and behaviors of traffic participants. The goal of interventions aimed at traffic participants is to reduce unsafe behaviors and traffic accidents or to promote traffic behaviors that reduce environmental impact and contribute to health promotion. To achieve these goals, it is necessary to communicate appropriately with the targeted group. Interventions targeting traffic participants can be considered as a type of risk communication and are based on research findings focusing on the cognitive and emotional processes of the message recipient. Theories and models of behavioral change in traffic, which encourage safer or more sustainable traffic behavior, often draw upon those used in the field of health behavior promotion. For more information on theories and models of behavioral change applicable to traffic behavior, please refer to further readings [38–41].

9.3.1 *Persuasive Communication*

In psychology, it is considered that in order to change behavior in individuals, it is necessary to first change their attitudes. The process of attitudinal change and persuasive communication as a method for achieving this has been studied extensively. Persuasive communications operate under the premise that people make rational

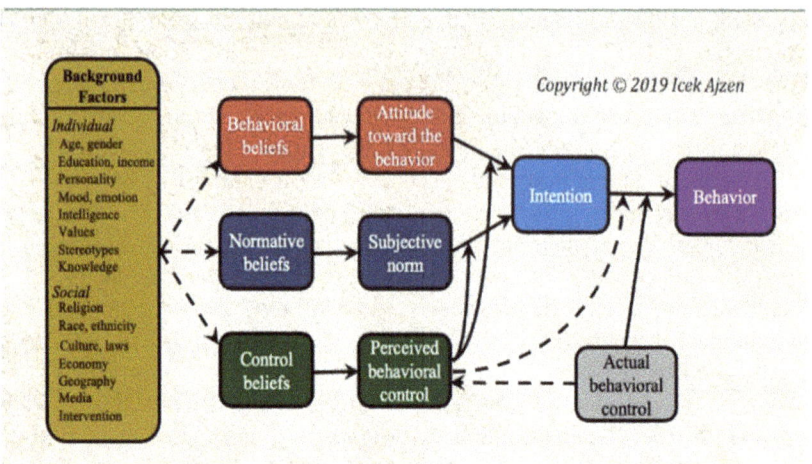

Fig. 9.5 Theory of planned behavior with background factors [42]

and reasonable choices after carefully considering the information given and their own situation. While many theories of persuasive communications have been proposed, this section will explain the Theory of Planned Behavior (TPB) [42] and the Protection Motivation Theory (PMT).

As shown in Fig. 9.5, TPB is a model that illustrates the relationship between "beliefs," "attitude," "intention," and "behavior," and has been widely applied in the study of traffic behavior. TPB originated from the Theory of Reasoned Action (TRA) proposed by Fishbein and Ajzen in 1975. TPB is a modification of the TRA that adds a "perceived behavioral control" to account for situations where behavioral control is not sufficient. TPB posits that the more favorable the attitude and subjective norms toward a behavior, and the greater the perceived behavioral control, the stronger the intention to perform the behavior [42]. Perceived behavioral control is a concept like self-efficacy, and it is assumed to be high when there is sufficient time, economic resources, and skills to take a certain action, making it easier to engage in that behavior. Based on TPB, to encourage individuals to engage in specific behaviors, interventions should aim to cultivate a positive attitude towards the behavior, demonstrate that the behavior is easy to perform, and highlight that significant others (important others) also support the individual's engagement in that behavior [41].

TPB has been applied in numerous studies of traffic behavior, with the majority focusing on explaining or predicting unsafe or high-risk behaviors of road users such as mobile phone use while driving, speeding, non-use of seatbelts, or traffic violations by pedestrians. Additionally, TPB-based questionnaires are often used to evaluate the effectiveness of advertising campaigns, safety education, and training interventions (for example, Box and Dorn [43]). It is important to note that merely referencing theories like TPB when developing interventions for advertising campaigns or safety education may not be sufficient. Separate procedures are necessary for planning and developing intervention programs.

Another example used for explaining and predicting traffic safety behavior is the Protection Motivation Theory (PMT) by Rogers (1975). PMT focuses on cognitive responses resulting from fear appeals [44]. Fear appeals are persuasive communications where the message sender emphasizes the dangers of a specific topic (and feelings of fear arising from it) to encourage the recipient's acceptance of coping behaviors to address that threat [45]. PMT posits that individuals engage in coping behaviors when they perceive a threat to a specific problem and are motivated to protect themselves from it. If a coping behavior is perceived as effective in avoiding the problem, and individuals believe they can perform it, and the implementation cost is low, then addressing fear appeals can effectively encourage the recipient to take action. However, if these conditions are not met, emphasizing the threat may reduce persuasive effectiveness. In relation to fear appeals, it is advisable not to overemphasize the threat if the recipient lacks sufficient information about coping behaviors or is not prepared for it, as this may diminish persuasive effects and can be counterproductive [41].

In traffic safety interventions, traditionally, efforts to induce attitudinal change have been made by addressing fear appeals, for example, by showing the tragic consequences of traffic accidents. In light of the principles mentioned above, education that only relies on fear appeals without providing concrete information about coping behaviors may not yield positive effects. Conversely, there are reports suggesting that education programs that carefully consider these aspects can lead to favorable effects in attitudinal change [46, 47].

9.3.2 Interventions Using Marketing Techniques

This section discusses two approaches to behavioral change using marketing techniques in broader contests, "nudging" and "social marketing." While persuasive communication theories such as TPB and PMT assume that individuals make decisions based on deliberation over available information and circumstances, there is a different venue for decision making. As discussed in Sect. 9.1.1 regarding the dual-process theory, there are two different types of thinking processes: System 1 and System 2. Similarly, in attitude research, a dual-process model of attitudes has been proposed, assuming two types of attitudes with different qualities, such as intuitive, context-dependent System 1, and deliberate System 2 [48]. In the dual-process model of attitudes, attitudes corresponding to System 1 are called "implicit attitudes," while those corresponding to System 2 are called "explicit attitudes." When studying habitual traffic behaviors or violations influenced by social desirability, it becomes more important to measure implicit attitudes. Thus, interventions aimed at behavioral change often focus on intuitive, context-dependent decision-making.

In part, behavioral economics emerged from an interest in people's decision-making based on System 1 of the dual-process theory [4]. One of its well-known applications is the technique called "nudging," which is utilized in various fields, including public policy. Nudging aims to minimize possible resistance that may be felt by individuals when trying to facilitate a new behavioral pattern [49]. Recent studies have presented practical application of nudges in traffic behavior, covering diverse topics such as acceptance of autonomous driving technology, reduction of driving speeds, and deterrence of pedestrian violation. To achieve the desired effects of nudging, implementers need to carefully consider the feasibility of adaptation in advance [50]. Furthermore, interventions based on nudges are essentially different from those based on education or training [49].

Social marketing is a comprehensive model encompassing theory and practice for behavior change, which has been practiced worldwide since Kotler and others proposed it in the 1970s [39, 51]. It refers to the process of planning, implementing, and evaluating programs to elicit voluntary actions from the target group by applying commercial marketing techniques for the public good, such as improving health, welfare, and safety. While there are similarities with the methods used in nudges, it is a different model. Differences from nudges include the emphasis on two-way communication between the intervention implementers and the target, the consideration of not only passive decision-making but also active decision-making by the target, and the preparation of not only incentives for the target but also disincentives for undesirable behavior choices. It is also recommended to systematically develop and implement programs using interdisciplinary teams, utilizing theories of behavioral science including persuasive communications [39, 51]. We may see an increase in research examples applying social marketing to traffic behavior studies in the future.

9.3.3 Interventions for Addressing Problematic Behaviors

When the problematic behavior of traffic participants is considered to pose danger to the safety of others and to themselves, special interventions need to be implemented to address the problems and prevent the recurrence of such behaviors. While nudges and social marketing interventions may not always be suitable for correcting problem behaviors, educational or therapeutic measures are necessary. Approaches applied to drivers' problematic behaviors include "cognitive behavioral therapy" and "motivational interviewing," or brief interventions simplified based on these psychotherapeutic models. Cognitive behavioral therapy is a short-term, structured psychological treatment based on the understanding that a person's mood and behavior are influenced by cognition (how things are thought about and perceived), and it aims to modify the way of cognitions and deal with problem [52]. Motivational interviewing is a semi-directive counseling approach that draws motivation from individuals who feel resistance to change, and guiding them to change their behavior on their own [53]. Interventions based on cognitive behavioral therapy or motivational interviewing are offered for drivers subjected to enforcement for drunk driving or caused traffic accidents (example reporting effectiveness [54]). Measures to prevent recidivism in drunk driving also require collaboration with medical experts because of alcohol dependency. Since many individuals with alcohol-related problems may not voluntarily seek treatment or education, motivational interviewing is used to overcome resistance to interventions (example reporting effectiveness [55]).

In recent years, there have been reports of application of cognitive behavioral model to reduce aggressive driving behaviors, commonly known as "road rage," characterized by hostile and threatening behaviors toward other road user [56, 57]. In Japan, programs aimed at preventing aggressive driving behaviors have been introduced in training sessions for individuals whose driver's licenses have been revoked [58].

9.4 Driver Assessment

This section discusses driver assessment. Driver assessment is conducted to provide appropriate interventions for drivers and is expected to play an important role in traffic safety measures. While referring to the system implemented in Japan as a "driver aptitude diagnosis," this section explores how to evaluate abilities and characteristics necessary for safe driving and further connect them to appropriate measures.

The importance of driver assessment is increasing in response to the growing prevalence of physical and mental health conditions among drivers and accidents caused by serious violations. In Japan, this corresponds to the system known as driver aptitude diagnosis. The concept of driving aptitude has its origin in the concept of "accident proneness" born in the early twentieth century. Accident proneness is explained as "personal traits that are relatively stable in individuals who experience more accidents than others." Stable traits include personality, intellectual ability, and psychomotor function. "Driving aptitude" refers to the characteristics of individuals that enable safe driving, but it is a concept that extends from accident proneness. Some countries, including Japan, have a system of driving aptitude diagnosis using psychological tests, but the definition of driving aptitude and methods of driver assessment vary widely depending on the country and region.

9.4.1 Accident Proneness

Accident proneness is a concept that has been the subject of debate among experts for many years. The background is detailed in previous studies [59–61]. Research on accident proneness flourished in Western countries from the time of industrialization and World War I until around the late 1960s. However, interest in accident proneness declined sharply thereafter. However, it does not mean that the existence of accident proneness itself—that some people are more likely to be involved in a crash—was denied. By that time, researchers realized that focusing solely on stable individual traits predisposing to accidents had limited utility in preventing traffic accidents and their consequences. Therefore, interest in accident proneness diminished rapidly. In fact, there are individuals who are more often involved in accidents or show unsafe behaviors, and psychological characteristics observed in such individuals have been reported [62, 63]. The drawback of solely focusing on accident proneness is the risk of attributing accidents solely to stable individual traits or specific human characteristics, thereby underestimating other possible measures such as driver education, training, improvements in driving environments, and technological advancements in safety measures. Accidents results from a variety of combinations of environmental as well as human factors, including not only personality traits associated with accident proneness but also temporal factors such as physical conditions, fatigue, use of substances, safety attitudes, motivation, lifestyle, factors that change over time such as age, driving experience, driving skills, and also socio-economic factors, gender, workplace, and family environment [60]. Therefore, instead of accident proneness, the use of a more neutral expression, "individual differences in accident involvement," [61, 64] has been proposed.

9.4.2 Driving Aptitude

Driving aptitude can be rephrased as "personal characteristics that enable safe and skilled driving of a car," [60] or simply, "the abilities necessary for safe driving" [65]. It is also a concept derived from the motivation to identify people who may be unfit for driving [66]. In Japan, research on driving aptitude began around the 1960s, leading to the development of various psychometric and psychomotor tests. They are widely utilized in driver training programs. In response to the sharp increase in drivers and traffic accident casualties at that time, the need to strengthen driver measures led to efforts in developing psychometric tests. In practice, driving aptitude tests are used as tools for education to encourage individuals to understand themselves and drive safely and not as means to identify unfit drivers.

In Japan, a range of psychometric tests are called driving aptitude tests that may be related to measurement and assessments of human factors related to safe driving. These tests encompass a wide range of evaluations, including assessments of cognitive function and screening tests, as well as measures of safe attitudes and those similar to hazard perception tests. This may be a unique practice in Japan. In essence, driving aptitude tests are aimed at measuring and evaluating individual differences in accident involvement or psychological characteristics of individuals that may lead to unsafe driving behaviors.

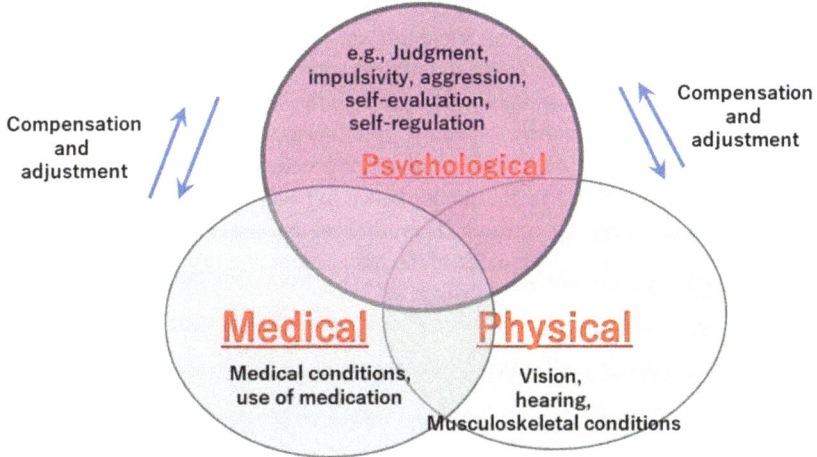

Fig. 9.6 Elements of driving aptitude (based on Okamura [67])

9.4.3 Driver Assessment in Japan and in Other Regions

As shown in Fig. 9.6, driving aptitude can be classified into medical, physical, and psychological elements [67]. These complement and adjust each other, and as a result, they are reflected in individual driving behavior. Among them, the role played by psychological elements is crucial and has a significant impact on driving behavior and accident involvement.

The term "driving aptitude" is not universally used worldwide. Globally, the expression "fitness to drive" is more common, and guidelines have been created in many countries regarding the medical and physical aspects of fitness to drive (for example, guidelines in the Great Britain [68]). On the other hand, detailed information about the concept and specialized techniques of psychological aspects of fitness to drive is often not publicly available. In this context, for example, Germany has built a driver assessment system that emphasizes psychological aspects without limiting it to medical aspects [65, 69]. In Germany, when a person who has had their driver's license revoked due to serious traffic violations such as repeated drunk driving wishes to regain their license, they must undergo a medical and psychological examination (Medizinisch-psychologische Untersuchung, MPU). The MPU is an assessment of drivers by medical and psychological experts to diagnose whether the lost driving aptitude has sufficiently recovered, based on detailed structured interviews and various test results. In the MPU, the psychological assessment plays a significant impact on the final diagnostic result. This is because whether the person who caused the accident or violations has sufficiently introspected and understood their problem and whether the risk of problem behavior can be sufficiently reduced are the criteria for diagnosis. A computer-based psychological tests equivalent to what is called a driving aptitude test in Japan is also utilized. However, the purpose is to confirm whether there are serious problems that cannot be compensated for by the individual, and if the minimum standards are met, it is judged as no problem. In other words, driving aptitude is not diagnosed solely on the basis of the results of psychological tests. Based on materials including the facts of the violation that triggered the revocation of the driver's license and the examinee's driving history, findings from interviews and observations of driving behaviors are also incorporated, and all information related to the person's driving behavior is utilized to make a holistic judgment of driving aptitude. Therefore, in driver assessment, it is essential to clarify the purpose and framework of the assessment and organize how to use the driving aptitude test.

References

1. Shinohara K (2011) Tyuui to human error: kotu anzen to tyuui mondai wo tyushin tosite. (Attention and human error: focusing on traffic safety and attention problems). In: Harada E, Shinohara K (eds) Modern cognitive psychology 4: attention and safety. Kitaoiji Shobo, pp 186–208
2. Rasmussen J (1986) Information processing and human-machine interaction: an approach to cognitive engineering. North-Holland
3. Stanovich KE, West RF (2000) Individual differences in reasoning: implications for the rationality debate? Behav Brain Sci 23:645–665. https://doi.org/10.1017/S0140525X00003435
4. Kahneman D (2011) Thinking, fast and slow, Farrar, Straus and Giroux
5. Keskinen E (2014) Education for older drivers in the future. IATSS Res 38:14–21. https://doi.org/10.1016/j.iatssr.2014.03.003
6. Renge K (2000) Psychological processes of risk-taking behavior in driving and new approach toward promoting risk-avoiding behavior. IATSS Rev 26:12–22
7. Wickens CD, McCarley JS, Gutzwiller RS (2023) Applied attention theory, 2nd edn. CRC Press/Routledge/Taylor & Francis Group, Routledge
8. Murphy G, Greene CM (2015) High perceptual load causes inattentional blindness and deafness in drivers. Vis Cogn 23:810–814. https://doi.org/10.1080/13506285.2015.1093245
9. Langham M, Hole G, Edwards J, O'Neil C (2002) An analysis of "looked but failed to see" accidents involving parked police vehicles. Ergonomics 45:167–185. https://doi.org/10.1080/00140130110115363
10. Miura T (2007) Shikakutekichuui no shinrigaku to kotsu anzen (Psychology of visual attention and traffic safety). In: Miura T, Harada E (eds) Psychology of accidents and safety - risk and human error. University of Tokyo Press, pp 129–155
11. Ball K, Owsley C, Sloane ME et al (1993) Visual attention problems as a predictor of vehicle crashes in older drivers. Invest Ophthalmol Vis Sci 34:3110–3123
12. Owsley C (1994) Vision and driving in the elderly. Optom Vis Sci 71:727–735. https://doi.org/10.1097/00006324-199412000-00002
13. National Center for Statistics and Analysis (2023) Distracted driving in 2021 (Research Note. Report No. DOT HS 813 443)
14. Stutts J, Feaganes J, Reinfurt D et al (2005) Driver's exposure to distractions in their natural driving environment. Accid Anal Prev 37:1093–1101. https://doi.org/10.1016/j.aap.2005.06.007
15. Masuda K, Haga S (2015) Effects of cell phone texting on attention, walking, and mental workload: comparison between the smartphone and the feature phone. Jpn J Ergon 51:52–61. https://doi.org/10.5100/jje.51.52

16. Horberry T, Osborne R, Young K (2019) Pedestrian smartphone distraction: prevalence and potential severity. Transp Res Part F Traffic Psychol Behav 60:515–523. https://doi.org/10.1016/j.trf.2018.11.011
17. Ishiai S (2003) Kouji nou kinou syougai gaku (Higher brain dysfunction). Ishiyaku Publishers Inc.
18. Miyake A, Friedman NP, Emerson MJ et al (2000) The unity and diversity of executive functions and their contributions to complex "frontal lobe" tasks: a latent variable analysis. Cogn Psychol 41:49–100. https://doi.org/10.1006/cogp.1999.0734
19. Adrian J, Moessinger M, Charles A, Postal V (2019) Exploring the contribution of executive functions to on-road driving performance during aging: a latent variable analysis. Accid Anal Prev 127:96–109. https://doi.org/10.1016/j.aap.2019.02.010
20. Daigneault G, Joly P, Frigon J-Y (2002) Executive functions in the evaluation of accident risk of older drivers. J Clin Exp Neuropsychol 24:221–238. https://doi.org/10.1076/jcen.24.2.221.993
21. Salthouse TA (1996) The processing-speed theory of adult age differences in cognition. Psychol Rev 103:403–428. https://doi.org/10.1037/0033-295X.103.3.403
22. Sit RA, Fisk AD (1999) Age-related performance in a multiple-task environment. Hum Factors 41:26–34. https://doi.org/10.1518/001872099779577345
23. Wechsler K, Drescher U, Janouch C et al (2018) Multitasking during simulated car driving: a comparison of young and older persons. Front Psychol 9. https://doi.org/10.3389/fpsyg.2018.00910
24. Matsuura T (2017) Kourei driver no anzen shinrigaku (Safety psychology for the older driver). University of Tokyo Press
25. Renge K, Mukai M, Ogawa K, Ota H (2007) Evaluation of the training of hazard perception for elderly drivers. IATSS Rev 32:6–13
26. Gugerty LJ (1997) Situation awareness during driving: explicit and implicit knowledge in dynamic spatial memory. J Exp Psychol Appl 3:42–66. https://doi.org/10.1037/1076-898X.3.1.42
27. Endsley MR (1988) Design and evaluation for situation awareness enhancement. Proc Hum Factors Soc Annu Meet 32:97–101. https://doi.org/10.1177/154193128803200221
28. Habibzadeh Omran Y, Sadeghi-Bazargani H, Yarmohammadian MH, Atighechian G (2023) Driving hazard perception tests: a systematic review. Bull Emerg trauma 11:51–68. https://doi.org/10.30476/BEAT.2023.95777.1370
29. Fukazawa N (1983) On the risk perception test (tentative name). Jpn J Appl Psychol 8:1–12
30. Ogawa K, Renge K, Nagayama Y (1993) A positive study on the structure and the function of hazard perception. Jpn J Appl Psychol 18:37–54
31. Shimazaki K, Mishina M, Nakamura A et al (2012) Driver education using a tablet device and movies of accidents recorded by drive recorders. Jpn J Traffic Psychol 28:35–43. https://doi.org/10.34362/jatp.28.35
32. Horswill MS, Marrington SA, McCullough CM et al (2008) The Hazard perception ability of older drivers. J Gerontol Ser B Psychol Sci Soc Sci 63:P212–P218. https://doi.org/10.1093/geronb/63.4.P212
33. Matsuura T (1999) Driver's overestimation of their own skill. Jpn Psychol Rev 42:419–437. https://doi.org/10.24602/sjpr.42.4_419
34. Zuckerman M (1994) Behavioral expressions and biosocial bases of sensation seeking. Cambridge University Press
35. Zhang X, Qu X, Tao D, Xue H (2019) The association between sensation seeking and driving outcomes: a systematic review and meta-analysis. Accid Anal Prev 123:222–234. https://doi.org/10.1016/j.aap.2018.11.023
36. Wilde GJS (1988) Risk homeostasis theory and traffic accidents: propositions, deductions and discussion of dissension in recent reactions. Ergonomics 31:441–468. https://doi.org/10.1080/00140138808966691
37. Wilde GJS (2001) Target risk 2: a new psychology of safety and health. PDE Publications

38. Conner M, Sparks P (2015) The theory of planned behaviour and the reasoned action approach. In: Conners M, Norman P (eds) Predicting health behaviour: research and practice with social cognition models, 3rd edn. Open University Press, pp 142–188
39. French J (2017) Social marketing and public health; theory and practice, 2nd edn. Oxford University Press
40. Glanz K, Rimer BK, Viswanath K (2015) Health behavior: theory, research, and practice, 5th edn. Jossey-Bass/Wiley
41. Imai Y (2006) Irai to settoku no shinrigaku: Hito ha tasya ni dou eikyou wo ataeruka (Psychology of request and persuasion: how to influence others). Selection Social Psychology-10. Saiensu-sha
42. Ajzen I (2019) Theory of planned behavior with background factors. https://people.umass.edu/aizen/tpb.background.html
43. Box E, Dorn L (2023) A cluster randomised controlled trial (cRCT) evaluation of a pre-driver education intervention using the theory of planned behaviour. Transp Res Part F Traffic Psychol Behav 94:379–397. https://doi.org/10.1016/j.trf.2023.03.001
44. Norman P, Boer H, Seydel ER et al (2015) Protection motivation theory. In: Conners M, Norman P (eds) Predicting and changing health behaviour: research and practice with social cognition models, 3rd edn. Open University Press, pp 70–106
45. Kimura K (2002) Kyoui ninchi, taisho ninchi to settoku: Bougo douki riron [Threat perception and coping perception and persuasion: protection motivation theory]. In: Fukada H (ed) Handbook of persuasion psychology. Kitaoji Shobo, pp 374–417
46. Cutello CA, Hellier E, Stander J, Hanoch Y (2020) Evaluating the effectiveness of a young driver-education intervention: Learn2Live. Transp Res Part F Traffic Psychol Behav 69:375–384. https://doi.org/10.1016/j.trf.2020.02.009
47. von Beesten S, Bresges A (2022) Effectiveness of road safety prevention in schools. Front Psychol 13. https://doi.org/10.3389/fpsyg.2022.1046403
48. Tosi JD, Haworth N, Díaz-Lázaro CM et al (2021) Implicit and explicit attitudes in transportation research: a literature review. Transp Res Part F Traffic Psychol Behav 77:87–101. https://doi.org/10.1016/j.trf.2020.12.014
49. Thaler RH, Sunstein CR (2021) Nudge: the final edition. Penguin Books
50. Rubaltelli E, Manicardi D, Orsini F et al (2021) How to nudge drivers to reduce speed: the case of the left-digit effect. Transp Res Part F Traffic Psychol Behav 78:259–266. https://doi.org/10.1016/j.trf.2021.02.018
51. Storey JD, Hess R, Saffitz G (2015) Social marketing. In: Glanz K, Rimer BK, Viswanath KV (eds) Health behaviour: theory, research, and practice, 5th edn. Jossey-Bass, pp 411–438
52. Ohno Y, Ito S (2011) Stress ya nigate to tsukiautameno ninchiryouhou: kituon tono tukiai wo toshite (Cognitive therapy and cognitive behavioral therapy for dealing with stress and weaknesses - through the relationship with stuttering). Kaneko Shobo
53. Harai H (2012) Houhou toshiteno doukizuke mensetsu (motivational interviewing as a method). Iwasaki Academic Publisher
54. Osilla KC, Paddock SM, McCullough CM et al (2019) Randomized clinical trial examining cognitive behavioral therapy for individuals with a first-time DUI offense. Alcohol Clin Exp Res 43:2222–2231. https://doi.org/10.1111/acer.14161
55. Ouimet MC, Dongier M, Di Leo I et al (2013) A randomized controlled trial of brief motivational interviewing in impaired driving recidivists: a 5-year follow-up of traffic offenses and crashes. Alcohol Clin Exp Res 37:1979–1985. https://doi.org/10.1111/acer.12180
56. Feng Z, Zhan J, Ma C et al (2018) Is cognitive intervention or forgiveness intervention more effective for the reduction of driving anger in Chinese bus drivers? Transp Res Part F Traffic Psychol Behav 55:101–113. https://doi.org/10.1016/j.trf.2018.02.039
57. Haustein S, Holgaard R, Ābele L et al (2021) A cognitive-behavioural intervention to reduce driving anger: evaluation based on a mixed-method approach. Accid Anal Prev 156:106144. https://doi.org/10.1016/j.aap.2021.106144

58. Traffic Bureau of the National Police Agency (2022) Torikeshi shobunsya koushu no unyou ni tsuite [On operation of the retraining course for those whose driving license was revoked] (notification). https://www.npa.go.jp/laws/notification/koutuu/menkyo/menkyo20221220_035.pdf
59. Haight FA (2001) Accident proneness: the history of an idea. Institute of Transportation Studies, University of California Irvine. Retrieved from https://escholarship.org/uc/item/9rh9f29x
60. Matsuura T (2000) Jikokeisei to untentekisei (accident propensity and driving aptitude). In: Renge K (ed) Social psychology of traffic behavior. Kitaoji Shobo, pp 18–26
61. Mckenna FP (1983) Accident proneness: a conceptual analysis. Accid Anal Prev 15:65–71. https://doi.org/10.1016/0001-4575(83)90008-8
62. Bowen L, Budden SL, Smith AP (2020) Factors underpinning unsafe driving: a systematic literature review of car drivers. Transp Res Part F Traffic Psychol Behav 72:184–210. https://doi.org/10.1016/j.trf.2020.04.008
63. Visser E, Pijl YJ, Stolk RP et al (2007) Accident proneness, does it exist? A review and meta-analysis. Accid Anal Prev 39:556–564. https://doi.org/10.1016/j.aap.2006.09.012
64. Elander J, West R, French D (1993) Behavioral correlates of individual differences in road-traffic crash risk: an examination of methods and findings. Psychol Bull 113:279–294. https://doi.org/10.1037/0033-2909.113.2.279
65. Gräcmann N, Albrecht M (2022) Assessment guidelines for driving suitability. In: Reports Fed Highw Res Institute Hum Saf. Issue M115. https://bast.opus.hbz-nrw.de/frontdoor/index/index/docId/2664
66. Yoshida S (1990) Problems of aptitude tests: thirty year development of SART research. IATSS Rev 16:249–258
67. Okamura K (2017) Unten tekisei (Driving aptitude). In: Matsuura T (ed) Series psychology and work. Traffic Psychology. Kitaoji Shobo
68. UK Driver and Vehicle Licensing Agency (2016) Assessing fitness to drive—a guide for medical professionals. https://www.gov.uk/government/publications/assessing-fitness-to-drive-a-guide-for-medical-professionals. Accessed 21 May 2024
69. Okamura K, Schimidt-Arndt S-B (2008) Measures for drink-drive offenders in Germany: assessment criteria of driving aptitude in medical-psychological examination. Jpn J Traffic Psychol 24:25–32

Open Access This chapter is licensed under the terms of the Creative Commons Attribution-NonCommercial-NoDerivatives 4.0 International License (http://creativecommons.org/licenses/by-nc-nd/4.0/), which permits any noncommercial use, sharing, distribution and reproduction in any medium or format, as long as you give appropriate credit to the original author(s) and the source, provide a link to the Creative Commons license and indicate if you modified the licensed material. You do not have permission under this license to share adapted material derived from this chapter or parts of it.

The images or other third party material in this chapter are included in the chapter's Creative Commons license, unless indicated otherwise in a credit line to the material. If material is not included in the chapter's Creative Commons license and your intended use is not permitted by statutory regulation or exceeds the permitted use, you will need to obtain permission directly from the copyright holder.

Chapter 10
Traffic Education

Yuto Kitamura, Kazuhisa Ogawa, and Nagahiro Yoshida

10.1 Introduction—"Traffic Education" as a Comprehensive Sciences

Education sciences related to traffic can be called "traffic education," but this term is not generally used. This is because it is difficult to say that the scientific system of education related to traffic has been sufficiently organized so far. Education sciences are academic approaches that verify educational contents, methods, policies, and systems from the standpoints of various disciplinary fields (for example, psychology, sociology, economics, etc.). The purpose of this chapter is to discuss the nature of traffic education as education sciences, focusing particularly on the importance of traffic safety education.

In this chapter, we want to define traffic education as thinking about the educational significance of traffic, which is the spatial movement of people and things by human will. In light of the above-mentioned nature of education sciences, traffic education is a field of study that verifies the educational contents, methods, policies, and systems that relate to traffic. One of its characteristics is that it relies on the knowledge of a wide range of academic fields.

For example, to think about how humans perceive others and other vehicles in the traffic space, and how they behave, it is indispensable to utilize various research results such as psychology, medicine, and physiology. In terms of learning about laws and social norms related to traffic, it is necessary to base on the knowledge of law, political science, sociology, etc. Furthermore, to understand how roads and

---

Y. Kitamura (✉)
Graduate School of Education, The University of Tokyo, Bunkyō, Japan
e-mail: yuto@p.u-tokyo.ac.jp

K. Ogawa
Center for General Education, Tohoku Institute of Technology, Sendai, Japan

N. Yoshida
Graduate School of Engineering, Osaka Metropolitan University, Osaka, Japan

© The Author(s) 2026
Pioneering the Future for Traffic and Safety Sciences,
https://doi.org/10.1007/978-981-96-0676-4_10

public transportation are maintained, knowledge of civil engineering and urban planning is also important.

In addition, traffic is closely related to people's lives and industries, so it has a deep connection with economics, geography, tourism, etc. Considering that the areas and ranges of people's interactions have expanded with the development of traffic, there is much to learn from the knowledge of history. In recent years, technological innovation has also advanced in the field of traffic, and traffic surveys using big data are conducted, increasing the importance of information science. Traffic education is a way of "learning" about various aspects of traffic while utilizing the research results and knowledge of a wide range of academic fields mentioned here.

Furthermore, we would like to point out that a characteristic of pedagogy is that it encompasses both normativity and empiricism. That is, when considering education, it is necessary to examine both the normativity of "should be" and the empiricism of "is." If we consider this in terms of traffic education, for example, when learning about traffic culture related to the joy of moving (i.e., the joy of travel or the fun of driving), normativity is emphasized, and when learning traffic regulations or acquiring appropriate driving skills, empiricism is emphasized.

Thus, traffic education is a "comprehensive sciences" that utilize knowledge from a wide range of fields, including humanities, social sciences, and natural sciences, to teach knowledge and skills about traffic and aims to nurture individuals who can perform appropriate traffic behavior. By nurturing such individuals, a safer traffic society is expected to be realized. Therefore, in this chapter, we would like to give an overview of the approach to traffic safety education within traffic education.

10.2 What Is Traffic Safety Education

Safety education refers to education aimed at developing the ability to predict and avoid dangers and the qualities and abilities to contribute to the safety of others and society. Such education is aimed at acquiring the necessary knowledge, gaining skills, and being able to act to ensure safety in various situations of daily life. In doing so, it is important to nurture the qualities and abilities to respect one's own and others' lives, lead a safe life throughout one's life, and contribute to the creation of a safe and secure society [1].

Safety education is divided into three areas: traffic safety, life safety, and disaster safety, and in each area, traffic safety education, crime prevention education, and disaster prevention education are practiced. Among these, in traffic safety education, it is required to "understand the dangers in various traffic situations and be able to walk safely and use bicycles and motorcycles" ([2], p. 29). In doing so, it is necessary to correctly judge information about safety on the road and ensure safety while paying attention to one's own physical and mental state and the way one behaves. For this purpose, it is required to "promote traffic safety education in a step-by-step and systematic manner" with the aim of spreading "traffic culture and traffic safety ideology" ([3], p. 10). In Japan, such traffic safety education is implemented through learning and activities in related subjects, such as moral education, integrated study time, and special activities (*Tokkatsu*), based on the Kindergarten Education Guidelines and the Course of Study for Elementary and Secondary Schools.

Also, it is important to consider traffic safety education that takes into account the environment surrounding children. If we do not consider the way of traffic safety according to the actual living environment, no matter how much appropriate traffic behavior training is conducted at school, it may result in not being able to protect oneself when actually falling into a dangerous situation. Therefore, when planning and implementing traffic safety education, it is necessary not only to collaborate with the local police but also to actively involve parents and local communities. Furthermore, it is also necessary to clearly position traffic safety issues in community development.

In addition to these points, we must not forget to plan and implement traffic safety education according to children's developmental stages and growth processes. The content of traffic safety education required varies according to the development of children's physical and mental health and changes in the living environment accompanying growth. For example, for elementary school students, the conditions vary depending on the situation of the school and the community. In Japan, the frequency of bicycle use increases around the fourth grade of elementary school, so it is necessary to transition from traffic safety education centered on walking behavior up to the third grade to those related to bicycle behavior for the fourth graders.

On the other hand, when they become junior high school students (i.e., the ninth graders), the school district becomes wider, and the circle of friends expands, so the range of behavior expands compared to when they were elementary school students; so we must consider traffic safety education that takes into account such behavioral changes. At that time, it is necessary to build a cooperative relationship beyond the school stage, such as collaboration between elementary and junior high schools, and cooperation between multiple schools within the school district (and in some cases, multiple schools beyond the school district). Along with these changes in children's behavior patterns, it is indispensable to plan and implement traffic safety education in accordance with the development of the minds and bodies of children.

Furthermore, when they become junior high and high school students, it is indispensable to deepen their understanding not only from the perspective of protecting themselves but also about the risk of becoming an offender, especially when driving a bicycle. Children themselves need to know that if an accident occurs, even minors may be held criminally and civilly responsible. For example, it is important to conduct traffic safety education that makes them think about their own responsibilities as members of society, taking into account cases where high compensation liabilities have occurred in bicycle accidents where teenagers were the offenders.

To realize such a broad range of traffic safety education, it is necessary to deepen the ties between schools, homes, and communities as mentioned above, and to consider improving the living environment and road conditions as a whole society. The importance of promoting comprehensive traffic safety education in schools is widely recognized, but it cannot be said that sufficient efforts are being made in schools in many countries (including Japan). This is because they are already spending a considerable amount of class time on school events in addition to regular classes, so there is no room to incorporate traffic safety education into the curriculum. Therefore, instead of trying to implement traffic safety education alone, we should consider incorporating traffic safety issues into cross-curricular and cross-domain learning through curriculum management.

10.3 Traffic Safety Education as Citizenship Education

Organizing what we have discussed so far, it becomes clear that it is important to consider traffic safety education from the three aspects of "head," "body," and "heart."[1] That is, "head" refers to the acquisition of knowledge about traffic safety. "Body" means practicing safe driving and walking behaviors, and being able to take appropriate actions and responses when encountering dangerous situations on the road or when unexpected situations occur. As for the "heart," it is important to raise awareness of traffic safety and cultivate a mentality that can coexist not only with oneself but also with others. We consider such education that nurtures these three aspects as citizenship education.

In order to introduce the approach of citizenship education to traffic safety education, it is necessary to adopt educational methods that encourage children to act autonomously on the road while having a public mind. This is realized as an educational practice that actively introduces problem-solving education and community-based activities, emphasizes participatory experiential learning, and learns about traffic safety in a cross-domain manner. For example, the European Commission's survey report [5] summarizing the efforts and challenges of traffic safety education in 25 EU countries would be a reference for such educational practices.

Furthermore, it is expected that children will function as mediators of "community building" through traffic safety education aimed at fostering citizenship and civic-mindedness. In other words, in traffic safety education that emphasizes the cultivation of children's civic and public nature, children are encouraged to think about what a safe and secure "community" should be like. At that time, it is necessary for various stakeholders in the community to think about how to create a safe and secure environment for children.

As already pointed out, when planning and implementing traffic safety education, it is required to utilize the knowledge of different professional fields and actively involve the people in the community. At that time, it is also important to

---

[1] The Ministry of Education, Culture, Sports, Science and Technology (MEXT) of Japan emphasizes the importance of nurturing "knowledge, morality, and physical strength" in a balanced manner as a characteristic of school education in Japan [4]. This concept also applies to traffic safety education.

position the school as one of the centers of community development. So far, we have overviewed traffic education, focusing particularly on traffic safety education, from the perspective of citizenship education.

In traffic safety education, it is not enough to just acquire knowledge about traffic laws and regulations and acquire skills to perform appropriate driving behavior. It is indispensable to cultivate attitudes and values that lead to safe traffic behavior. As Ogawa [6] points out, "Traffic safety education is not something that can be completed with just one learning. It is important to provide traffic safety education as a spiral curriculum, a process that develops various learning experiences in a spiral manner" (p. 300). For example, when considering bicycle driving behavior, it starts with acquiring vehicle operation skills, and then it is necessary to cultivate the ability to judge whether there is any danger in the traffic situation. In this way, after forming the basis for safe behavior, enhance skills to consider driving purposes and contexts, such as choosing a safe route or refraining from unnecessary bicycle use during late-night hours.

Acquiring the "life purpose and skills for living" that enable a safe and comfortable traffic life while controlling motives and attitudes that take high risks and are impulsive, such as speeding or ignoring signals, which increase the possibility of causing accidents, is set as the highest educational goal [6]. In order to conduct such traffic safety education, we would like to further consider what to pay attention to and how to proceed.

10.4 Theoretical Framework for Implementing Traffic Safety Education at Schools

10.4.1 *How to Develop What Kind of Qualities and Abilities*

The "Third Plan for the Promotion of School Safety" presented by the Ministry of Education, Culture, Sports, Science and Technology [7] states that the goal of safety education is "for all students to acquire the qualities and abilities related to safety so that they can make appropriate judgments and take initiative in their actions." In recent school education, from the perspective of "what can be achieved," it is aimed to develop the qualities and abilities needed in the new era. Based on this direction, the current curriculum guidelines classify the important qualities and abilities that support "the power to live" into three pillars: "knowledge and skills," "thinking power, judgment power, expressive power, etc.," and "the power to learn, humanity, etc.," and aim to develop the qualities and abilities corresponding to each pillar [8]. In order to promote traffic safety education at schools, it is necessary to concretize the content and methods of education, considering the relationship with the direction shown in these curriculum guidelines. In doing so, by incorporating the following three perspectives into the content and methods of education, more effective traffic safety education is expected to be realized.

10.4.2 Perspective 1: Cultivating Qualities and Abilities Related to Risk Management and Crisis Management

Safety education at schools is not limited to the field of traffic safety. School safety consists of three areas: "traffic safety," "life safety (including crime prevention)," and "disaster safety (synonymous with disaster prevention)," and traffic safety education in school education is one activity positioned within these three areas [9]. In educational activities conducted both inside and outside school facilities, including commuting to and from school, various situations can occur that threaten the lives of students. For example, situations where students are involved in accidents or disasters, such as falling into heat stroke or cardiac arrest during club activities, incidents related to crime prevention occurring during commuting, being forced to evacuate due to the occurrence of earthquakes or fires, and traffic accidents during commuting, can be assumed. In fact, there are countless cases where students are involved in such accidents and disasters, resulting in death or serious injury.

On the one hand, the school education timetable is overcrowded, making it difficult to secure the number of hours allocated for safety education (currently, safety education is not positioned as a subject in the Japanese curriculum, but is often implemented as an activity in areas such as special activities, comprehensive learning (inquiry) time, as a unit of subjects such as health and social studies, or as a cross-curricular activity). Therefore, it is required to effectively and efficiently promote safety education in the three areas, and it is necessary to set a common framework for learning between the areas. In this book, the purpose of safety education is defined as "cultivating the qualities and abilities necessary to adapt to the road traffic environment (living environment, natural environment) where the risk of traffic accidents (life accidents, natural disasters) lurks."

By defining it in this way, even if the environment to be adapted changes, if the qualities and abilities related to "Risk Management" and "Crisis Management" are necessary as basic and generic forces, the commonality of learning can be achieved. Predicting and choosing avoidance actions for potential dangerous situations in the future, minimizing damage even when faced with a crisis, there are many commonalities in the process of thinking and judgment to be learned in the three areas of safety education. Also, the learning outcomes of traffic safety education can be applied to life safety and disaster safety, promoting the transfer and generalization of learning, and if the framework of learning is common, it is possible to effectively and efficiently promote safety education in the three areas.

After all, we live in a continuum of various risk scenes such as traffic accidents, life accidents, crime incidents, natural disasters, etc. If we demand different frameworks for each area, we will demand complex thinking and judgment from children, and there is a risk of confusion in adapting to the environment. The three areas of school safety are set by adults, and children do not originally distinguish them, and there is no need to divide the three areas and learn them. For example, by presenting a picture of a school route and asking, "What kind of danger can be predicted? How can we stay safe? Please think from the perspective of traffic safety (crime prevention, disaster prevention)", just by switching the perspective, there is a method to learn risk management and crisis management in the three areas at once, which is not a particularly difficult task for children.

Of course, from the perspective of safety management, it is first necessary for the adults around them (teachers, community members, guardians, etc.) to engage in risk management and crisis management activities to make the environment in which children live safer. On the other hand, children themselves also need to minimize the risk of accidents and disasters they (or others) are involved in as much as possible, and even if they face an accident or disaster, they need to acquire the ability to minimize the damage, which ultimately leads to protecting their own and others' lives. For example, activities such as creating a traffic safety (crime prevention, disaster prevention) map of the school district and learning to predict and avoid dangers using photos of the school route are effective educational activities where students themselves proactively learn the qualities and abilities related to risk management and crisis management, such as where and what kind of dangers are lurking, how to act to commute safely, and how to minimize damage even if faced with a crisis. Especially, the creation of safety maps is an educational activity that can be practiced in the framework of learning common to three areas, and it has been introduced in many schools.

10.4.3 Perspective 2: Associating with the Components of Qualities and Abilities Necessary for Safe Behavior

So, specifically, what content should be included in traffic safety education? The answer to this question is derived from the answer to the question of what elements our safe behavior is composed of. In recent years, the qualities and abilities required for safe driving have a four-tiered structure, and the idea of designing a curriculum while considering the systematic connection of the content included in each tier is widely accepted in the field of driver education. This idea is called the "Hierarchical Approach" to driver behavior. The GDE model introduced later (Sect. 10.5.2) [10] is based on the idea of the hierarchical approach, which organizes and classifies educational goals and content corresponding to each tier.

In the school setting, when implementing traffic safety education for students, considering that the school routes and school hours are relatively limited, it is thought that knowledge and skills related to driving planning (strategic level in the GDE model) can be included and aggregated in other tiers, and this book proposes a hierarchical approach assuming a three-tier structure of qualities and abilities. To make it easier to introduce into current school education, the layers corresponding to the "operational level," "tactical level," and "meta level" from the lower layers of the GDE model are modeled as "basic road safety and participation in traffic," "adaptation to road traffic environment," and "contribution and responsibility to local safety".

The lower level of the GDE model, "Operational Level (Qualifications & Abilities necessary for vehicle operation)," can correspond to skills such as knowledge about traffic rules and road usage, how to walk on roads, how to ride a bicycle, etc., in the case of commuting to school. It is thought that participation in the traffic society is possible only with the acquisition of basic knowledge and skills, and this Level 1 Qualifications & Abilities is named "Basic Safe Road Use and Participation in Traffic." The next Level 2, "Tactical Level (Qualifications & Abilities necessary for adapting to driving situations)," includes cognitive processes such as thinking, judgment, and decision-making of action selection, such as hazard prediction and hazard avoidance, and is thought to correspond to Qualifications & Abilities related to adapting to constantly changing situations. This Level 2 is named "Adaptation to Road Traffic Environment." And the upper level, "Meta Level (Qualifications & Abilities related to lifestyle)," is considered to correspond to citizenship education. Understanding how one's actions affect others and learning content about contributing and responsible actions as a citizen are included in this layer, so Level 3 is named "Contribution and Responsibility to Local Safety."

By structuring it in this three-layered manner, we can roughly align it with the three pillars of "vitality for life" introduced earlier (Sect. 10.4.1)—namely, "knowledge and skills," "thinking ability, judgment, and expressive power," and "the ability to learn and human qualities." This approach also ensures consistency with the curriculum guidelines. In "Basic and Participation in Traffic for Safe Road Use," the focus is on learning basic knowledge and skills for participating in road traffic, and in "Adaptation to Road Traffic Environment," the learning content includes cognition (thinking), decision-making (judgment), and action selection (expression) in situations where accident risks lurk. The relationship between "Contribution and Responsibility to Regional Safety" and "The Power to Learn, Humanity, etc." can be found in the perspective of considering one's role in safety in relation to society and learning a better way of life as a citizen.

10.4.4 Perspective 3: Consider the Developmental Stages of Students

What meaning would the mental structure of qualifications and abilities being hierarchical have when planning systematic learning? One is the direction of learning, which is the accumulation from the lower layers. To participate in a traffic society, it is necessary to first master the basics, starting with understanding traffic rules such as the meaning of signals. However, the actual road traffic environment is complex, and even if traffic rules and regulations are followed, safety cannot be fully ensured. For example, when crossing an intersection at a green light, there is a risk of intersecting with right-turning vehicles from the front or rear, and in addition to obeying signals, it is necessary to constantly check the surrounding situation and make appropriate situation judgments and action selections. Therefore, the second layer of educational content is necessary.

Once students have acquired the ability to protect their own lives, we want them to acquire the ability to ensure the safety of others. It is important to develop qualifications and abilities related to risk management and crisis management for the lives of others, such as being able to take actions to ensure the safety of elderly people and people with disabilities in the community, and being able to demonstrate exemplary behavior to young children and juniors. In other words, it is a systematic path to develop learning from "Basic and Participation in Traffic for Safe Road Use" to "Adaptation to Road Traffic Environment" and then to "Contribution and Responsibility to Regional Safety."

In developing learning in this direction, it is important to consider the developmental stages of students. According to Piaget's theory of cognitive development, preschoolers and lower elementary school students are in the preoperational stage, and are influenced by egocentrism, making it difficult for them to think from an objective perspective when perceiving their environment. Children of this age group often run out into the street. From the child's perspective, a play space is formed between them and their friend on the other side of the road, and they are simply moving within that space, without the perception that they are running out into the road. Therefore, instead of instructing them not to run out, it is more effective to encourage behavior improvement by using positive expressions such as "stop," "look," and "check" when entering the road, from the child's perspective. Thus, for preschoolers and lower elementary school students, the focus should be on acquiring the basics of safe behavior by focusing on the first layer, "Basic safe road use and participation in traffic." As children reach the middle grades of elementary school, they enter the concrete operational stage and are able to objectively perceive their surrounding environment. They can also predict the appearance of approaching vehicles from behind obstacles at intersections with poor visibility and can logically recognize danger without being influenced by appearances. From this age group, it is desirable to start learning educational content corresponding to the second layer, "Adaptation to the road traffic environment." Specifically, this includes creating traffic safety maps, and learning to predict and avoid danger using photos and videos. For example, presenting a photo of an intersection and asking open-ended questions such as "What kind of danger can be predicted when crossing?" and "How can you cross safely?" and sharing the students' opinions with the whole class are effective in developing environmental adaptability. For middle and high school students in adolescence, including content related to "Contributing to and taking responsibility for safety in the community" in educational activities, can enhance students' interest from a new perspective. Distributing traffic safety maps created by the students to elderly people in the community, instructing new students on safe bicycle commuting, and visiting nearby elementary schools and kindergartens to give traffic safety lectures are all ways in which students can actively participate in educational activities, enhancing their self-involvement in traffic safety issues and facilitating attitude change and behavior improvement. It is important to expand and develop risk management and crisis management awareness from personal safety to the safety of others and the community according to the developmental stages of students and to increase the weight of the educational content of the upper layers as they grow. On the other hand, once the qualities and abilities of the upper layers have been somewhat acquired, it is also acceptable to relearn the content included in the lower layers. After thinking about the safety of the community, you may reflect on your own behavior, such as whether your bicycle riding is appropriate, and this may improve your basic behavior to be safer.

10.5 Traffic Behavior Models Related to Traffic Safety Education

Generally, humans are not born with skills or abilities related to safe traffic behavior, and some people may engage in insufficiently safe traffic behavior or unintentionally/intentionally incorrect traffic behavior. In the Safe System Approach, which was born against this backdrop, human vulnerability is recognized, just as it is a fact that humans make mistakes and do not necessarily obey rules. In addition, children and adolescents may be at risk due to physical and psychological constraints. Factors that explain or predict such specific traffic behaviors involve physiological, psychological, social, and cultural aspects, so it is important to understand these background factors correctly and consider interventions to promote changes in traffic behavior. Therefore, in this section, we introduce some models that explain the relationships between human safe traffic behavior and background factors, which can be applied to traffic safety education, from psychological/social perspectives.

10.5.1 Basic Model of Traffic Safety Education

The purpose of traffic safety education is to positively influence safe traffic behavior. To achieve this goal, it is important to focus on human psychological factors and (1) deepen knowledge and understanding of traffic rules and traffic situations, (2) improve skills through training and experience, and (3) strengthen and/or change attitudes toward danger, personal safety, and the safety of other road users [11]. This concept is shown in Fig. 10.1.

For example, it is expected that by acquiring knowledge of traffic signs and traffic rules through education, one will be able to choose correct traffic behavior in accordance with traffic rules. In addition to knowledge, there are skills related to vehicle operation and dynamic ability, and attitudes, which are psychological readiness related to traffic behavior, and these can be considered as internal factors that influence traffic behavior.

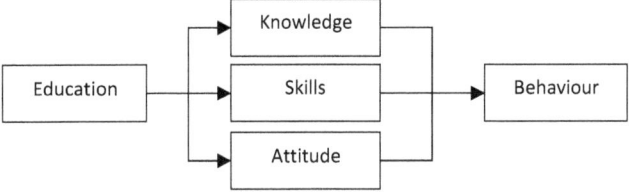

Fig. 10.1 Fundamental model of road safety education

In Japan, for example, the Traffic Safety Education Guidelines, which show the concept of traffic safety education based on this basic model, specifically indicate the required knowledge, skills, and attitudes according to age, means of transportation, and characteristics of those receiving traffic safety education.

10.5.2 Training Model

A training model based on risk conditions and learning processes for individual drivers operating vehicles and other vehicles is the Goals for Driver Education (GDE) matrix. This is a representation of the driver's tasks as a hierarchical schema, consisting of different control levels (Table 10.1).

Here, a schema is a framework for understanding limited information from the outside world, which humans acquire through the accumulation of experience. From the bottom of the table, it consists of four levels: the vehicle control level, which includes driving skills; the tactical level, which includes driving adjustments under changing traffic conditions; the strategic level, which considers the goals and background factors behind driving; and the meta-level, which considers driver attitudes, decision-making and motivations, etc.

The higher levels are directly related to the lower levels. The higher levels express that there is a direct relationship with the lower levels. Within each level, toward the side of the table, it consists of basic knowledge and skills required while driving, risk-increasing factors that drivers should be aware of, and self-assessment, which deals with how accurately drivers assess their own abilities.

In driver training based on this model, it is considered that emphasizing driving tasks at one higher level and considering educational programs contribute to the formation of more essential and safer drivers, i.e., people who can take safe traffic behavior on their own. Although the model was developed for drivers, it can be applied to other traffic participants, and the pillar of the approach is to coach learners to understand how their personalities, values, and beliefs influence the way they drive.

Table 10.1 Goals of driver education model

Hierarchical level of behavior (extent of generalization)	Content of driver education		
	Knowledge and skills the driver has to master	Risk-increasing factors the driver must be aware of and be able to avoid	Self-evaluation
Goals for life and skills for living (global)	Knowledge about/control over how general life goals and values, behavioral style, group norms, etc. affect driving	Knowledge about/control over risks connected with life goals and values, behavioral style, social pressure, substance abuse, etc.	Awareness of personal tendencies—impulse control, motives, lifestyle, values, etc.
Goals and context of driving (specific trip)	Knowledge and skills—trip-related considerations (effect of goals, environment choice, effects of social pressure, evaluation of necessity, etc.).	Knowledge and skills—Risks connected with trip goals, driving state, social pressure, purpose of driving, etc.	Awareness of personal planning skills, typical driving goals, driving motives, etc.
Mastery of traffic situations (specific situation)	General knowledge and skills—rules, speed adjustment, safety margins, signaling, etc.	Knowledge and skills—wrong speed, narrow safety margins, neglect of rules, difficult driving conditions, vulnerable road-users, etc.	Awareness of personal skills, driving style, hazard perception, etc., from the viewpoint of strengths and weaknesses
Vehicle maneuvering (specific situation)	Basic knowledge and skills—maneuvering, vehicle properties, friction, etc.	Knowledge and skills—risks connected with maneuvering, vehicle properties, friction, etc.	Awareness of personal strengths and weaknesses—basic driving skills, maneuvering in hazardous situations, etc.

Hatakka et al. [10]; Engström et al. [12]; Siegrist [13]

10.5.3 The Theory of Planned Behavior

A model widely used in the field of traffic psychology is the "Theory of Planned Behavior (TPB)," which focuses on the fact that human behavior is determined by internal factors [14]. This theoretical model emphasizes the existence of an intention to act and posits that this intention to act is influenced not only by attitudes but also by perceived control and subjective norms. It also takes into account cases where not all factors that affect the actual performance of human behavior can be controlled. There is also a rational action approach, which is a successor to the theory of planned behavior, that has evolved the hypothesis that self-efficacy is important in determining the strength of an individual's intention to act (Fig. 10.2).

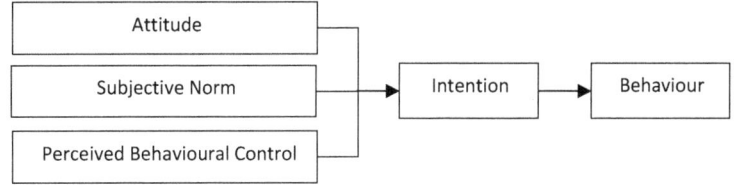

Fig. 10.2 Theory of planned behavior [15]

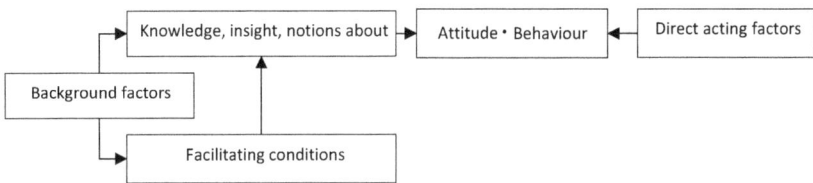

Fig. 10.3 A comprehensive model of behavior modification [16]

10.5.4 Behavior Modification Model

The behavior modification model was originally introduced to change various health behaviors and is an approach to change behavior using operant conditioning. Here, operant conditioning targets spontaneously occurring behaviors, and by conditioning with rewards or punishments, it is expected to increase or decrease spontaneous behaviors and thus bring about behavior modification. In this situation, where the stimulus-response relationship is examined, it is possible to consider not only the attributes and social situations of the subjects assumed to be the place of traffic safety education but also the internal factors that want to be changed (Fig. 10.3).

10.5.5 Approaches Targeting Traffic Behavior

In the field of behavioral science, there are models that assume various internal factors. In the field of traffic safety, the definitions of desirable/undesirable behaviors are diverse, and it is not always clear how interventions, including education, play a role. In order to gain a better understanding of the relationships between these various related elements, we introduce a framework that incorporates the concept of behavior choice in Fig. 10.4.

The procedure for considering interventions along this framework is as follows: (1) understand the background factors of the situation where undesirable behavior is chosen and (2) concretize desirable behavior and set goals. After these goals are clearly set, (3) it is necessary to consider interventions related to background factors such as education, infrastructure, and enforcement. Here, when considering

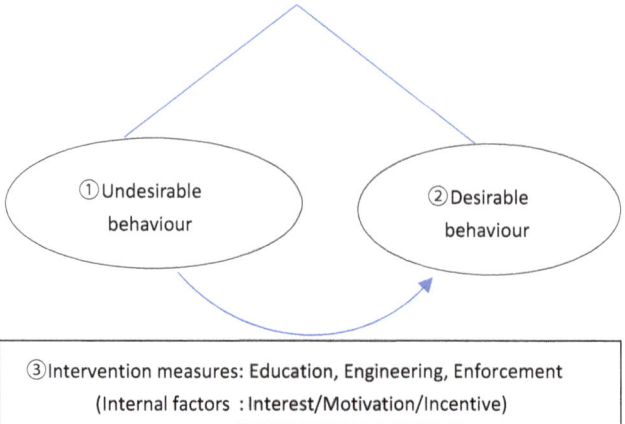

Fig. 10.4 A framework that incorporates the concept of choice of transport behavior

interventions, it is necessary to explicitly deal with the process of change in human internal factors and to stimulate them, and it is important to capture the triggers of behavior change, such as increasing interest/motivating/providing incentives, as important elements.

10.6 Conclusion

In this chapter, after explaining the outline of traffic education as a comprehensive science, we introduced the details of traffic safety education, which is a particularly important area covered by traffic education. Traffic safety education is indispensable for people to acquire appropriate knowledge and skills about traffic and to actually take safe traffic actions and behaviors. Especially for children who are vulnerable road users and young people who can be both victims and perpetrators as active road users, it is required to acquire the necessary knowledge, skills, and attitudes through traffic safety education. However, in many countries, including Japan, it is hard to say that traffic safety education is sufficiently practiced in school education. In order for traffic safety education to be actively practiced, it is essential that such problem awareness is widely shared in society.

At the same time, in order to practice and develop effective traffic safety education, it is also necessary to understand its theoretical foundation. Therefore, aiming to be a reference for those involved in traffic safety education, this chapter introduced the theoretical framework of traffic safety education and the main models that explain the relationship between human safe traffic behavior and background factors. Of course, the theories and models presented here are only a part, so it is indispensable to actively utilize the knowledge of various fields of traffic education as comprehensive sciences. We hope that traffic education, which is composed of various academic fields, will develop across fields; that the knowledge generated there will be applied to further enrich the content, methods, policies, and systems of traffic education; and that a safer and more secure traffic society will be realized. We would like to conclude this chapter with this hope.

References

1. Ministry of Education (2022) Culture, sports, science and technology of Japan (MEXT). Third Plan for the Promotion of School Safety
2. Ministry of Education, Culture, Sports, Science and Technology of Japan (MEXT) (2019) Safety education at schools that Foster "Zest for Living"
3. Ihara K, Masaoka T (eds) (2011) The significance and role of traffic safety education - based on the activities of the Kagawa prefecture traffic safety education promotion council. Keisō Shobō
4. Ministry of Education, Culture, Sports, Science and Technology of Japan (MEXT) (Central Council for Education) (2008) Improvement of the course of study for kindergartens, elementary schools. Junior High Schools, High Schools and Special Needs Schools
5. Rose 25 (2005) Inventory and Compiling of a European Good Practice Guide on Road Safety Education Targeted at Young People (Final Report). Rose 25 (Project funded by the European Commission)
6. Ogawa K (2007) Evaluation study of traffic safety education program for children - 'dangerous spot mapping'. IATSS Rev 32(4):299–308
7. Ministry of Education, Culture, Sports, Science and Technology (2022) The 3rd plan for the promotion of school safety
8. Ministry of Education, Culture, Sports, Science and Technology (2016) Improvement of the Course of Study for Kindergartens, Elementary Schools, Junior High Schools, High Schools, and Special Needs Schools and Necessary Measures (Report)
9. Ministry of Education, Culture, Sports, Science and Technology, School Safety Materials (2019) Safety education at schools that foster the power to live
10. Hatakka M, Keskinen E, Gregersen NP, Glad A, Hernetkoski K (2002) From control of the vehicle to personal self-control: broadening the perspectives to driver education. Transp Res Part F 5:201–215
11. Böcher W (1995a) Traffic enlightenment and traffic education. In: Hilse H-G, Schneider W (eds) Traffic safety. Handbook for the development of concepts. Boorberg, Stuttgart u.a, pp 248–299
12. Engström I, Gregersen NP, Hernetkoski K, Keskinen E, Nyberg A (2003) Young novice driver education and training, literature review, VTI-rapport 491A. Swedish National Road and Transport Research Institute, Linköping
13. Siegrist S (1999) Driver training, testing and licensing: towards theorybased management of young drivers' injury risk in road traffic; results of EU-Project GADGET, work package 3. Schweizerische Beratungsstelle für Unfallverhütung, bfu
14. Ajzen I (1985) From intentions to actions: a theory of planned behavior. In: Kuhl J, Beckmann J (eds) Action control, SSSP Springer Series in Social Psychology. Springer, Berlin, Heidelberg
15. Böcher W (1995b) Verkehrsaufklärung und Verkehrserziehung/Traffic Safety Campaigns and Traffic Safety Education. In: Hilse H-G, Schneider W (eds) Verkehrssicherheit. Handbuch zur Entwicklung von Konzepten/Traffic Safety. Manual for Development of Concepts. Boorberg, Stuttgart u.a, pp 248–299
16. Bjørnskau T, et al (2017) The Norwegian Council for Road Safety's model for behaviour modification

Open Access This chapter is licensed under the terms of the Creative Commons Attribution-NonCommercial-NoDerivatives 4.0 International License (http://creativecommons.org/licenses/by-nc-nd/4.0/), which permits any noncommercial use, sharing, distribution and reproduction in any medium or format, as long as you give appropriate credit to the original author(s) and the source, provide a link to the Creative Commons license and indicate if you modified the licensed material. You do not have permission under this license to share adapted material derived from this chapter or parts of it.

The images or other third party material in this chapter are included in the chapter's Creative Commons license, unless indicated otherwise in a credit line to the material. If material is not included in the chapter's Creative Commons license and your intended use is not permitted by statutory regulation or exceeds the permitted use, you will need to obtain permission directly from the copyright holder.

Chapter 11
Traffic Safety and Medicine

Kazuhiko Kibayashi, Takashi Moriya, Migiwa Asano, and Masaya Takahashi

11.1 Traffic Accident Casualty Statistics

The National Police Agency's traffic accident casualty statistics cover accidents involving vehicles, trams, and trains on roads as defined by the Road Traffic Act, which result in death or injury. The number of people who die within 24 h of an accident is counted, and this number is generally referred to as the number of traffic accident fatalities. To capture those who die after 24 h have passed since the accident and to make international comparisons, since 1993, the number of people who die between 24 h and 30 days after the accident has also been counted in addition to the 24-h fatalities. As for injuries, cases requiring treatment for 30 days or more are classified as serious injuries, and those requiring treatment for less than 30 days are classified as minor injuries, and the number of injuries is the total of serious and minor injuries [1]. From a medical perspective, all deaths involving transportation, regardless of the location of the accident or the time after the accident, are considered traffic accident deaths.

The number of traffic accident fatalities (24-h fatalities) nationwide has been decreasing almost every year, with 11,452 in 1992 as the peak, 7768 in 2003, 4117 in 2015, and 2610 in 2022. The number of people who die between 24 h and

---

K. Kibayashi (✉)
Department of Forensic Medicine, Tokyo Women's Medical University, Tokyo, Japan
e-mail: kibayashi.kazuhiko@twmu.ac.jp

T. Moriya
Department of Emergency and Critical Care Medicine, Saitama Medical Center, Jichi Medical University, Saitama, Japan

M. Asano
Department of Legal Medicine, Graduate School of Medicine, Ehime University, Shitsukawa, Toon, Ehime, Japan

M. Takahashi
National Institute of Occupational Safety and Health, Kawasaki, Japan

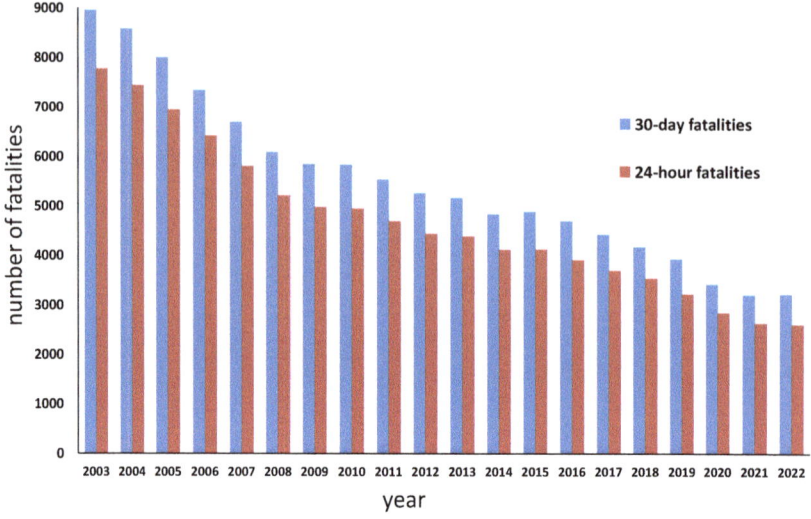

Fig. 11.1 Number of traffic fatalities

30 days after an accident has also been decreasing almost every year, with 1176 in 2003, 768 in 2015, and 606 in 2022, and 14–23% of the number of people who die within 24 h of an accident die between 24 h and 30 days after the accident (Fig. 11.1). The number of injuries has also been decreasing almost every year, with 1,181,681 in 2003, 666,023 in 2015, and 356,601 in 2022, and serious injuries requiring treatment for 30 days or more account for about 6–8% of the injuries each year (Fig. 11.2). The decrease in the number of casualties is the result of comprehensive accident prevention measures such as stricter penalties for drunk driving, mandatory seat belt use, improvements in road environments that make accidents less likely and vehicle structure development, improvements in public transportation and emergency medical systems, and traffic safety education.

When the number of traffic accident fatalities is broken down by age group, the number of fatalities among the elderly aged 65 and over is high, especially among those aged 75 and over. Although the number of fatalities in each age group is decreasing almost every year, the rate of decrease in the number of elderly fatalities is small, so the proportion of fatalities among the elderly aged 75 and over in the total number of fatalities has been increasing, from 26% in 2006 to 36% in 2015 and 39% in 2022 (Fig. 11.3). In 2022, the proportion of traffic accident casualties (fatalities and injuries) among the elderly aged 75 and over, totaling 26,202, was highest for those in cars at 36%, followed by pedestrians at 30% and cyclists at 28%. In addition to cognitive function tests for elderly drivers, elderly driver courses, and driving skill tests, measures for elderly pedestrians and cyclists are also important

11 Traffic Safety and Medicine

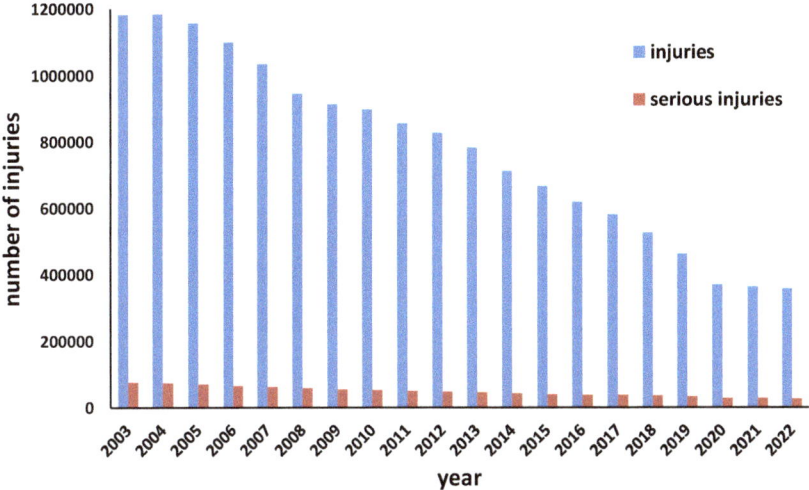

Fig. 11.2 Number of injuries in traffic accidents

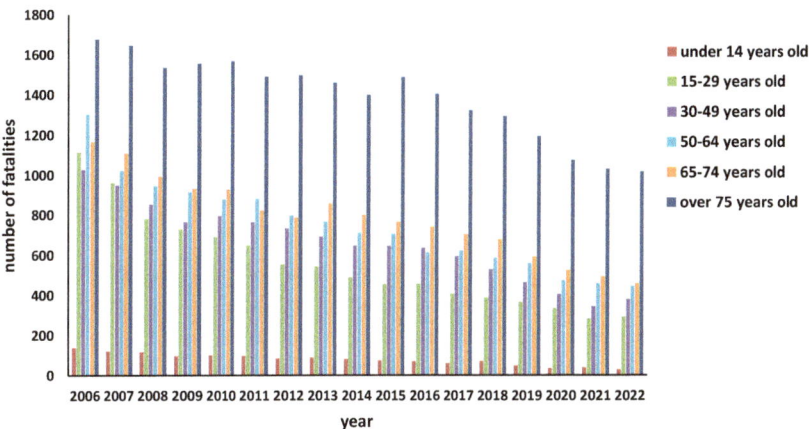

Fig. 11.3 Number of traffic fatalities by age group (fatalities within 24 h)

(Fig. 11.4). The "11th Basic Plan for Traffic Safety" announced by the Cabinet Office aims to reduce the number of deaths within 24 h to less than 2000 (less than 2400 deaths within 30 days) and the number of seriously injured to less than 22,000 by 2025. In addition to the police statistics on deaths due to traffic accidents, there are also demographic statistics based on death certificates. These traffic accident statistics serve as a basis for understanding the trends in traffic accidents and further reducing the number of casualties and realizing a safe traffic society [2].

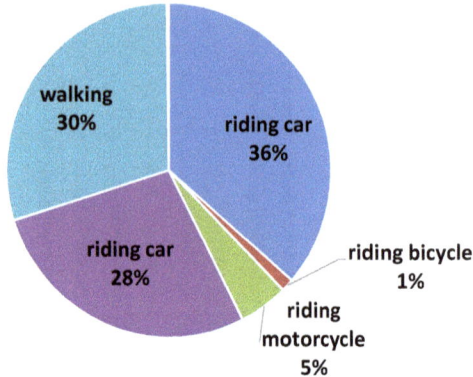

Fig. 11.4 Number of casualties by type of traffic accident (75 years and older, 26,202 persons, 2022)

11.2 Traffic Accident Injuries

Traffic accidents cause damage to the human body, and severe damage can lead to death. In all deaths within 24 h due to traffic accidents, the most common fatal injury site is the head/face, accounting for about 40%, followed by the chest (Fig. 11.5). In all serious injuries requiring more than 30 days of treatment due to traffic accidents, the most severe injury site is the limbs, accounting for about 40%, followed by the head/face, and chest (Fig. 11.6). Brain injuries among head/face injuries are frequent and severe, and research on diagnostic methods and pathophysiological analysis of brain injuries is an important issue that leads to patient survival [3, 4]. In cases of life-threatening head injuries, in addition to preventing the accident itself, reducing injuries through measures such as wearing helmets is important for further reducing the number of deaths.

Victims of automobile traffic accidents are classified into pedestrians, drivers, and passengers, with other categories including victims of accidents involving motorcycles and bicycles. Pedestrian injuries are categorized into those caused by the initial collision with the vehicle (primary injuries), those caused by being thrown up and colliding with the vehicle (secondary injuries), and those caused by being slammed onto the road surface (tertiary injuries). Pedestrian injuries are formed depending on the shape of the vehicle and the speed at the time of collision. Injuries to drivers and passengers occur when they collide with the interior structure of the vehicle when it collides with other vehicles or roadside structures and also occur during vehicle rollovers or when occupants are ejected from the vehicle. Serious injuries such as heart rupture can occur in steering wheel injuries when the driver's chest and abdomen collide with the steering wheel. Seat belts prevent the head and chest and abdomen from colliding with the front windshield or steering wheel, and also prevent ejection from the vehicle, which can cause serious injuries. Seat belt use is mandatory not only for drivers and front seat passengers but also for rear seat passengers, and the spread of seat belt use is reducing the number of severe injuries and deaths among drivers and passengers. Seat belt use is also necessary when airbags are deployed, and if the seat belt is not worn, the rapidly inflated airbag can strike the chest and cause serious injuries.

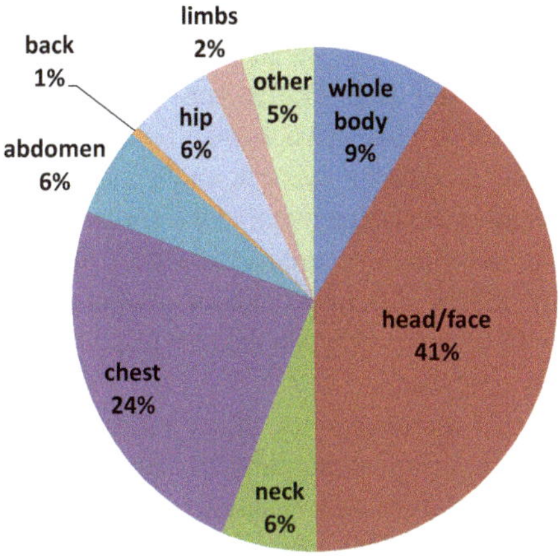

11 Traffic Safety and Medicine

Fig. 11.5 Number of fatalities by site of injury (2610 deaths within 24 h of accident, 2022)

Fig. 11.6 Number of seriously injured persons by site of injury (26,027 persons, 2022)

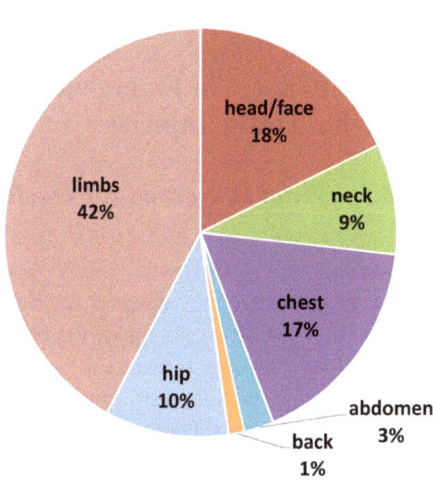

Bicycles are vehicles that can be easily used by everyone from children to the elderly, but bicycles are one of the means of transporting people and goods, and accidents involving bicycles are traffic accidents. Traffic accidents involving bicycles include ① collisions between bicycles and cars, ② collisions between bicycles and pedestrians, ③ collisions between bicycles, and ④ single bicycle accidents. In intersections where bicycles collide with cars, it is common for the bicycle to fall over and the rider to suffer serious head injuries. In collisions between bicycles and pedestrians, or between bicycles, pedestrians, or bicycle riders may also suffer serious head injuries due to falling. In single bicycle accidents, the rider may fall into roadside ditches with the bicycle, resulting in spinal cord injuries or drowning. Bicycle riders need to obey signals at intersections, obey stop signs, and check for safety. When passing near pedestrians, they need to slow down, and sometimes it is necessary to get off the bicycle and push it past. Attention is also needed for accidents involving infants falling from child seats on bicycles [5].

The number of bicycle traffic accidents has been increasing, and many of the fatalities from these accidents have suffered fatal injuries to the head. As a result, the revised Road Traffic Act was enacted on April 1, 2023, stipulating that "bicycle operators must strive to wear helmets for riding." The obligation for bicycle users to make an effort to wear helmets has been extended from those under 13 years old to all ages. It is expected that the spread of helmet use among bicycle users will reduce the number of severe head injuries in case of falls. However, it should be noted that the effect of helmets is hard to expect in cases where pedestrians fall in collisions with bicycles or when the cervical cord is damaged by falling into a side ditch alone on a bicycle.

The analysis of the injury mechanism in traffic accidents is basic data for ensuring the safety of people, vehicles, and road environments to prevent traffic accidents and casualties (active safety) and to reduce the damage to occupants and pedestrians after an accident (passive safety) [6].

11.3 Advances in Emergency Medical Care for Traffic Injuries

11.3.1 What Is Needed for Doctors Involved in Trauma Care?

Diseases treated in emergency medicine can be divided into endogenous diseases, illnesses, and exogenous diseases, as well as injuries. Endogenous means that the cause of the disease occurs inside the body. If the blood vessels in the brain fail due to high blood pressure, it results in cerebral hemorrhage; if the blood vessels in the heart are obstructed due to arteriosclerosis induced by cholesterol, it results in acute myocardial infarction. These are typical examples of endogenous diseases, and both are called killer diseases because they lead to death if left untreated. On the other hand, exogenous diseases are characterized by causing diseases due to external factors, and traffic injuries are their leading cause. Heatstroke caused by a hot environment and drug poisoning caused by taking a large amount of drugs are also classified as exogenous. These can also lead to death if left untreated. Still, it was revealed in the 1990s that there are cases where people die even if they are transported to the hospital by ambulance because the procedures for trauma care and resuscitation are insufficient. There are cases where the patient died unexpectedly from trauma (preventable trauma death: PTD), even though the probability of death would not have been high if daily trauma care had been provided, it is defined, and although it has been said that a certain proportion exists, it was a number that greatly exceeded expectations, which is still fresh in the memory [7]. This situation was 20 years behind the United States, where trauma care is advanced. In endogenous diseases, it is common to hear the patient's main complaint, conduct an examination, make a diagnosis from the test results, and then start treatment. The origin of such medical practices is Western medicine, ingrained in medical students. However, in trauma care for exogenous diseases, the basic practice is to examine the overall condition in the order of airway, breathing, circulation, and consciousness before paying attention to the patient's complaints and to improve the overall condition from abnormal findings of physiological signs (respiration rate, pulse, blood pressure, body temperature, consciousness). After stabilizing the overall condition, the local damage is treated. If the state of consciousness is not clear, there is a tendency to prioritize head CT, and it is also necessary to be careful not to be distracted by visible trauma, which can be a pitfall in trauma care. In response to these characteristic trauma treatments, the Japan Trauma Care Research Organization has developed the JATEC (Japan Advanced Trauma Evaluation and Care) course, which continues to educate mainly doctors in emergency care at emergency medical centers nationwide through off-the-job training courses. Similarly, the course is also being expanded for nurses. Furthermore, numerous clinical studies have been reported the based on nationwide registration of cases by JTDB (Japan Trauma Data Bank) [8, 9].

It is a well-known social issue that the number of ambulance requests is increasing in a aging society, consistent with the increase in endogenous diseases. On the other hand, exogenous diseases are decreasing due to the spread of crisis management and safety education. However, since the primary victims are young people, the social loss is significant, and there are cases where they cannot live with their conditions before the injury due to sequelae. Unexpected death is an essential concept for improving trauma care, but we must also respond to unexpected trauma disability (preventable trauma disability) in the future.

11.3.2 Development of Pre-hospital Medical Care and Rescue

Doctor helicopter and doctor car services are being deployed nationwide. It is crucial to start hemostasis or surgery within an hour of trauma, as it can affect the prognosis. This is the most important time for those who perform trauma treatment. Dispatching doctors and nurses to the scene as quickly as possible and starting treatment after evaluating the patient at the scene or in the ambulance is very beneficial. Recently, attention has been paid to diseases such as acute myocardial infarction and ischemic cerebrovascular disease, which emphasize time. The key for these disease groups is to recognize the disease and establish a system to reduce the transport time [10]. It is important to transport patients efficiently and accurately to a trauma hospital with trauma care staff on trauma, as the time of a traffic accident is generally clear. As of April 2022, 56 doctor helicopters have been deployed in all 47 prefectures, including Kyoto Prefecture, which belongs to the Kansai Wide Area Union, which operates doctor helicopters. However, there is currently no base hospital in Kyoto Prefecture. Doctor helicopters cannot fly in bad weather or at night, but they can transport over distances that would take more than an hour by ambulance in a few minutes. This mobility and widespread nature are leading to the concentration of severe trauma cases. The issue is who and how to recognize

Fig. 11.7 The doctor's car is traveling to the scene of a traffic accident

severely injured patients and decide on the operation of the doctor's helicopter. However, each emergency medical technician in the medical control department plays a role. Doctor cars (Fig. 11.7) can operate 24 h a day and has advantages in urban areas. The system for transporting doctors and nurses to the scene early not only shortens the time to start treatment but is also thought to be connected to the medical system of the transport organization, based on the information of the treating doctor.

11.3.3 Progress of Medical Care in Hospitals

A paradigm shift that can advance emergency medical care is being realized by starting treatment with a doctor's helicopter or doctor's car, performing resuscitation treatment, and transporting the patient to a medical institution where appropriate treatment can be performed (Fig. 11.8).

The doctor's treatment in the ambulance improves the overall condition, making safe and reliable trauma treatment possible. Recently, the first modality in Japan, called Hybrid ER (Emergency Room), has been introduced and is attracting attention (Fig. 11.9).

This is a room that has been made all-in-one with a CT scan to examine the whole body for injuries, an angiography device to check for bleeding points, embolization to treat bleeding points, and an operating room function for severe cases that cannot be transported to a general operating room, all around the examination table to enable rapid and safe treatment in severe trauma care. It is possible to enter the hybrid ER while stabilizing the overall condition in the ambulance and carring out resuscitation treatment and fundamental treatment. Instead of waiting for severely ill patients in the hospital, doctors and nurses go out of the hospital. In other words, securing one treatment space outside the hospital can be considered a paradigm shift. The hybrid ER has drawbacks such as high introduction costs and

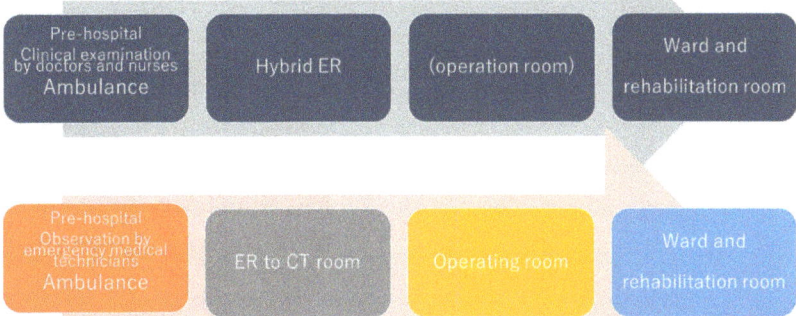

Fig. 11.8 A paradigm shift for faster initiation of treatment in emergency medical care

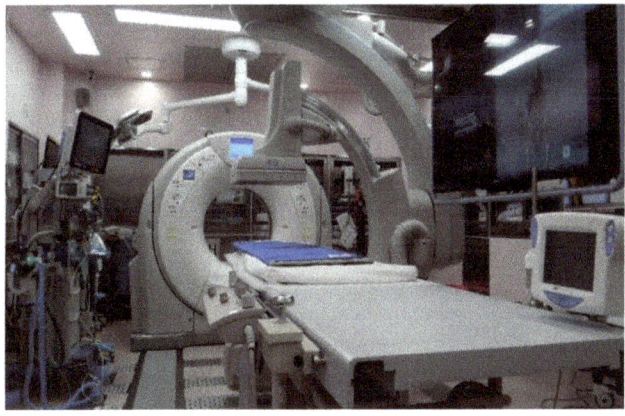

Fig. 11.9 HERS (hybrid emergency room system) in Jichi Medical University, Saitama Medical Center

difficulty in handling a large number of patients; however, it is also true that scattered reports have emerged of saving severely ill patients who could not be saved before.

11.4 Sudden Death while Driving

Sudden, unexpected internal death (sudden death) was defined as death from the disease within 24 h of onset. Sudden death can occur in any situation, such as during sleep, work, or meals. However, when it occurs while driving a car, it is a problem

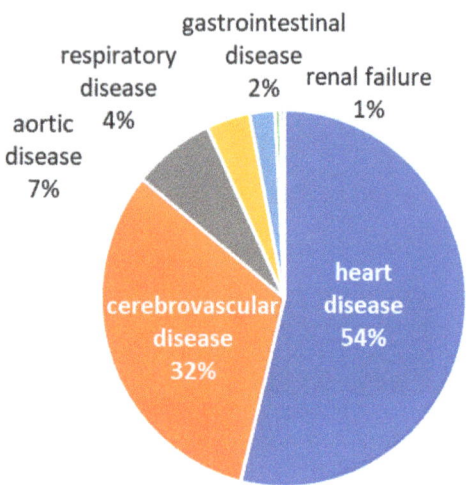

Fig. 11.10 Sudden death while driving a motor vehicle. (Based on Ref. [13])

not only for the death of the driver but also for humans and physical damage caused by collateral accidents.

It has been reported that 8.3% [11] of car drivers involved in traffic accidents and 9% [12] of car collisions is caused by diseases. The primary causes of sudden death while driving is cardiovascular disease, such as ischemic heart disease, cerebrovascular disease, and aortic disease [12, 13]. In an analysis report for 50 years from 1953 to 2003 at the Tokyo Medical Examiner's Office, where all unnatural deaths in the target area were examined or autopsied by medical examiners to diagnose the cause of death [13], sudden death while driving a car accounted for 71.5% of those aged 40–64, with heart disease accounting for 53.4%, followed by cerebrovascular disorders at 31.8% (Fig. 11.10). The rate is high for ordinary passenger cars and taxis. Sudden death from ischemic heart disease while driving is common in men aged 40–60 with a history of hypertension, myocardial infarction, and fatty liver, all of whom have stenosis of the coronary artery; 57.1% died shortly after the onset, and the remaining 42.9% died within 24 h after the onset of myocardial infarction [14]. In an analysis of professional taxi drivers [15], half of those who were unable to continue driving due to some disease died shortly after that; the causes of death were cerebrovascular disease (42.9%), heart disease (30.0%), and aortic disease (7.1%), as previously reported. Recently, accidents while driving has increased among older people in Japan, and the actual number of sudden deaths while driving may be higher than previously reported [11].

In cases of sudden death while driving, it is of great concern whether the cause of death is a disease or trauma from a traffic accident. There are reports [16] that no external injuries were observed in cases of sudden death while driving; however, this was based on an analysis of only four cases. Instead, in forensic autopsy cases, it is not uncommon for a victim to sustain trauma associated with a traffic accident, even if the cause of death is a disease. According to an analysis of 46 cases of sudden death while driving where a forensic autopsy was performed, 66.9% had injuries that were thought to have been caused by a traffic accident but were not the cause of death [17]. In our cases of sudden death while driving, the collision speed was often relatively low (approximately 40 km/h), and external injuries were relatively minor. It is presumed that the loss of consciousness preceding the collision caused the accelerator pedal to be released, making the steering wheel and brake operations impossible.

Without an autopsy, it is often difficult to determine whether a death was caused by trauma or illness. This affects mortality statistics and is closely related to the presence or absence of negligence and eligibility for insurance payments. Therefore, it should be diagnosed correctly from a medical perspective. Although an autopsy is a meaningful means of diagnosing the cause of death, the autopsy rate for unexpected and unnatural deaths in Japan is low. In areas without a medical examiner system, the cause of death is diagnosed without performing an autopsy for most unnatural deaths, excluding criminal corpses and unusual deaths. However, the spread of postmortem imaging diagnosis is remarkable, and in Japan, where the autopsy rate is low, CT imaging is performed at the hospital to which the body is transported immediately before or after death, or at the time of autopsy, and it helps in diagnosing the cause of death. The correct diagnostic rate of traumatic death by postmortem CT examination is high [18], and the diagnostic accuracy of CT examination for traumatic death due to traffic accidents has significantly improved [19]. Postmortem image searches should be conducted to investigate the causes of death.

In cases where ischemic heart disease developed while driving and was transported to the hospital, there were scattered cases that showed lethal arrhythmias, such as ventricular fibrillation and ventricular tachycardia. Most sudden deaths while driving occur in a state of cardiac arrest at the time of contact with the emergency team; however, the time from the sudden changes in physical condition to the discovery of the accident is short, and it has been pointed out that if life-saving measures are taken by bystanders immediately after the accident, death can be avoided [20].

Both professional drivers and the public should be required to undergo health examinations. When older people renew their licenses, it may be worthwhile to conduct risk assessments for ischemic and cerebrovascular disease and test for dementia and driving ability. However, preventing sudden death complex, making it even more challenging to prevent sudden death while driving. In addition, it is essential to prevent collateral accidents and save drivers' lives. It is hoped that a car control system will be developed to detect sudden driving incapacity, quickly notify emergency medical institutions in the event of a sudden illness, and detect sudden changes in the driver to evacuate or safely stop the vehicle.

11.5 Commercial Motor Vehicle Crashes Related to Medical Conditions

Commercial motor vehicle (CMV) crashes related to medical conditions in our country are defined as "cases where the driver can no longer continue to operate a CMV due to illness" [21]. More specifically, they can be considered as CMV crashes caused by the sudden worsening of a diagnosed pre-existing disease or the acute onset of an undiagnosed disease. The target diseases are mainly cerebrovascular disease, cardiovascular disease, sleep apnea syndrome (SAS), and visual impairment, but diabetes, epilepsy, mental illness, sleep disorders other than SAS, and dementia are included as well [22]. It is not easy to predict that such diseases will cause CMV crashes at the start of driving. However, once such a crash occurs, significant damage is not unavoidable for the health and lives of not only the driver, but also passengers, pedestrians, and people around.

11.5.1 Current Situation of CMV Crashes Related to Medical Conditions

When the number of CMV crashes related to medical conditions is expressed per 10,000 drivers working in each transport type of trucks, taxis, chartered/courtesy buses, and transit buses, this ratio for the transit buses has been outstandingly high in recent years (Fig. 11.11). However, most of them were interruptions of duty, and cases leading to collisions or contacts are few. In the remaining transport types, the ratio for chartered/courtesy buses was high until 2019, and that of trucks and taxis was gradually increasing. Nevertheless, for the trucks and taxis, we have to pay attention to many cases leading to personal injury accidents or property damage accidents accompanied by collisions or contacts [21].

The vertical axis represents the number of CMV crashes per one-million drivers. The right panel depicts only the data for truck, taxi, and charterd/courtesy bus for clarity.

11.5.2 Efforts to Prevent CMV Crashes Related to Medical Conditions

By the authorities' effort, various manuals and guidelines for preventing these crashes are available. When the awareness of these documents was investigated for employers, over 80% of the employers for truck, bus, and taxi drivers were aware of

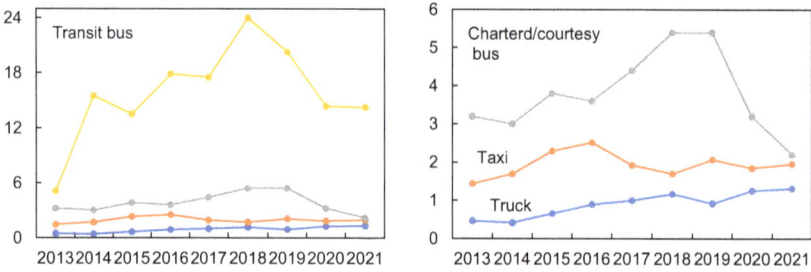

Fig. 11.11 Yearly trend of commercial motor vehicle (CMV) crashes related to medical conditions by transport type

Fig. 11.12 Percentage of employers conducting a screening test by disease and transport type (FY2022)

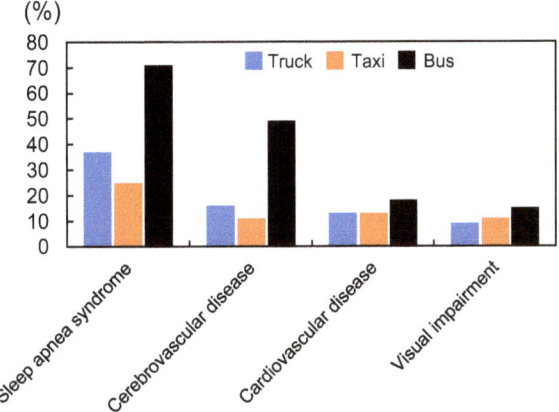

the documents about health management, SAS, cerebrovascular disease, and cardiovascular disease [21]. The manual for visual impairment measures was recently published (in 2022), and its awareness was between 60% and 80%. Screenings for SAS, cerebrovascular disease, cardiovascular disease, and visual impairment are being conducted, but according to a survey of the employers, the significant difference was found depending on the type of transport and disease [21] (Fig. 11.12). While screening can identify groups with high driving impairment associated with each disease, the cost of medical examination and the response after the diagnosis are major issues in the workplace.

Considering why such sudden changes in physical condition occur, not only high-risk approaches such as screening, but also health management, labor management, and self-management across all drivers become important. Drivers must undergo a health checkup once or twice a year. If abnormal findings indicated, the employer must receive industrial physician's opinions and also take appropriate measures for working conditions to maintain the health of the driver. Furthermore, at each roll call, blood pressure measurement/verification and sleep state examination are required [23]. Unfortunately, it is a great challenge for the authorities to expand such occupational health services into small-size workplaces that make up the majority of the transportation industry.

Driving work involves heavy demand, including long hours and night shift work, and is closely related to overwork-related death and disorders (cerebrovascular/cardiovascular heart disease and mental disorders) [24, 25]. Regulations on the upper limit of overtime work for drivers started as of April 2024 in Japan, and we need to ensure a balance between work and health [26].

In the medium to long term, it is essential to install systems that can stop or evacuate vehicles to the shoulder in response to changes in the driver's physical condition. In addition, engineering technologies to detect abnormal driving or contact and stop safely will also be taken.

The CMV crashes caused by health impairment are nothing but a tragedy for everyone. In order to maximize the various advantages of moving by car, efforts to minimize the risk of this sort of crashes are indispensable.

11.6 Drinking, Medication, and Driving

Alcohol, antianxiety drugs, sleeping pills, psychotropic drugs, dangerous drugs, stimulants, marijuana, and other illegal drugs have pharmacological actions that interfere with normal driving and pose a risk of serious traffic accidents. Among foreign countries, in France, alcohol is detected in 2.1% of drivers and marijuana in 3.4% of drivers [27], and the most frequently detected in traffic accident participants internationally are marijuana and alcohol. However, in Japan, alcohol is overwhelmingly the majority.

In Japan, around 2011, drugs with synthetic cannabinoids and cathinone, as their main ingredients spread as quasi-legal drugs and herbs. Traffic accidents caused by these drugs have rapidly increased and become a social problem. However, by comprehensively regulating these dangerous drugs and strengthening enforcement, the number of traffic accidents caused by dangerous drugs significantly decreased from approximately 2015 onward. Various studies [28–30] show that illegal drugs, such as marijuana and dangerous drugs, increase the risk of traffic accidents and the severity of injuries when used in combination with alcohol and sleeping pills and that the use of multiple types of drugs is hazardous.

In contrast, antidepressants and antihistamines are thought to cause drowsiness and delay reaction time because of their pharmacological effects, but the risk of accidents while driving is not as high as that of alcohol or illegal drugs [31].

11.6.1 Alcohol (Ethanol)

Alcohol is a beverage and, at the same time, a drug with central nervous system inhibitory effects. Drinking is involved in 5–35% of traffic accident deaths [32], making alcohol a critical risk factor for traffic accidents. The probability of a driver under the influence of alcohol causing a fatal accident is 17.8 times higher than that of a sober driver [27]. In Japan, the number of drinking-related deaths and serious accidents is decreasing (Fig. 11.13), but by 2022, fatal accidents caused by drinking will account for 5.5% of all traffic fatalities, approximately seven times higher than non-drinking fatalities, and the rate is still high [33].

Alcohol is a hydrophilic, low-low-molecular-weight compound; therefore, when taken orally, it is quickly absorbed from the digestive tract and diffuses into water without binding to proteins. Approximately 95% of absorbed alcohol is metabolized in the liver in two stages. It is oxidized by alcohol dehydrogenase to acetaldehyde, oxidized by aldehyde dehydrogenase (ALDH) to produce acetic acid, and finally

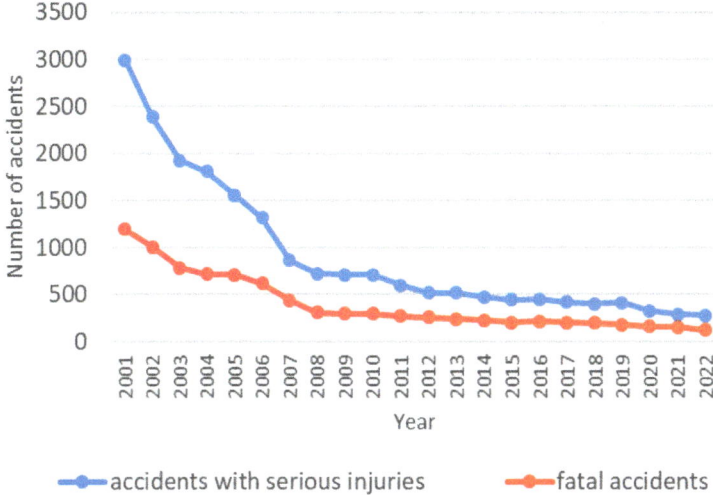

Fig. 11.13 State of traffic accidents caused by drunk-driving in Japan [33]

excreted from the body as carbon dioxide and water. Only a small amount of ingested alcohol is excreted unchanged, including in urine, feces, breath, and sweat. The breath alcohol test conducted in traffic investigations measures alcohol excreted in the breath unchanged. The concentration of alcohol in the breath correlates with the concentration of alcohol in the blood, and the concentration in the breath is approximately 1/2000.

Genetic polymorphism has also been identified in ALDH. While the normal type of ALDH is common in non-Asians, including Westerners, the deficient type is common in East Asians, and approximately 40% of Japanese people carry the inactive ALDH2*2 allele [34]. Individuals with ALDH deficiency have slower alcohol metabolism, resulting in differences in blood alcohol concentrations and perceived symptoms of intoxication after drinking.

Alcohol can lead to a decrease in driving ability and an extension of reaction time; the higher the blood alcohol concentration, the higher the risk of traffic accidents [35]. However, even in small amounts, this can lead to decrease attention and judgment, negatively affecting driving [36]. Interestingly, the symptoms of intoxication and blood concentrations do not always match. Generally, the feeling of intoxication is strongly apparent during a rise in blood alcohol concentration, and during a decline, even if the blood alcohol concentration is still high, one often feels much sober than in the previous state [37]. Additionally, subjective feelings of intoxication vary significantly among individuals. Individuals with low activity of acetaldehyde dehydrogenase (ALDH deficiency type) feel sober when their blood alcohol concentration drops close to zero. In contrast, those with high activity (ALDH normal type) feel completely sober during the early stage of the decline in blood alcohol concentration, that is when the blood alcohol concentration is still considerably high [38].

Generally, the blood alcohol disappearance rate for people of the ALDH normal type is 0.16–0.20 mg/mL/h, and for the ALDH deficiency type, it is 0.14–0.15 mg/mL/h [37]. Therefore, reducing the blood alcohol concentration (1.6 mg/mL), which causes slurred speech and unsteady movement (staggering) to zero, would take more than 8 h, even for people with normal ALDH. In Japan, it is rational that the Aviation Law prohibits pilots and others from drinking alcohol 8 h before duty. A predriving alcohol test is useful to avoid driving a car without realizing that one is drinking. Previously, aviation workers were required to take alcohol tests before their work. However, since 2011, automobile drivers have been required to use an alcohol detector before and after their duty to check for alcohol consumption. Since then, the number of traffic accidents caused by alcohol has decreased [1]. Furthermore, in 2022, the obligation for alcohol testing was expanded by non-transportation business operators to eradicate drinking/driving accidents. The number of drink-driving accidents is expected to decrease in the future.

11.6.2 Dangerous Drugs

Among abused drugs, synthetic cannabinoids and cathinone compounds are considered "dangerous drugs" in Japan and are globally known as and are collectively referred to as new psychoactive substances (NPS). Cannabis and synthetic cannabinoids are agonists of the central cannabinoid CB1 receptor, and their effects can be observed within seconds to minutes of inhalation. Symptoms of intoxication include nausea, vomiting, euphoria, happiness, changes in sensation, including sharpening of hearing and vision, panic attacks, and effects on the cardiovascular system, causing tachycardia and fainting. Symptoms that can cause traffic accidents include changes in the spatial perception of sensations. There are typical cases where the road deviates owing to a mistake in the steering operation because the road appears curved despite being straight with good visibility [39]. Cathinone compounds exhibit stimulant and cocaine-like central excitatory effects, as well as euphoria. In Finland, it has been reported that the cathinone compound methylenedioxypyrovalerone (MDPV) was detected in 5.7% of drivers under the influence of drugs, suggesting that cathinone compounds increase the risk of traffic accidents [40]. In addition, the combined use of MDPV, stimulants, and benzodiazepines [40] is common, and it is clear that this has a serious impact on traffic safety.

References

1. National Public Safety Commission and National Police Agency. Statistics about Road Traffic. Portal site of official statistics of Japan (e-Stat); 2023
2. Shimada R, Kibayashi K (2022) Changes in the number of traffic collisions during the various waves of COVID-19 infection in Japan. PLoS One 17(12):e0278941

3. Kibayashi K, Shimada R, Nakao KI (2021) Analysis of autopsy cases involving individuals who experienced cardiopulmonary arrest immediately after sustaining minor head injuries. J Forensic Leg Med 81:102205
4. Tatara Y, Shimada R, Kibayashi K (2021) Effects of preexisting diabetes mellitus on the severity of traumatic brain injury. J Neurotrauma 38(7):886–902
5. Kibayashi K, Ezaki J, Hasegawa M (2016) Analysis and prevention of deaths caused by bicycle-related traffic accidents. IATSS Review 41(2):106–113
6. Kibayashi K (2019) Prevention of head trauma and death in patients with head injuries: a forensic autopsy study. IATSS Res 43(2):71–74
7. Otomo Y, Henmi H, Honma M, Masuko K, Kosaki K, Yokota J, Murata A, Shimazaki S (2002) Quality evaluation of medical care at the Accepting Hospital is essential for selecting medical facilities for severe trauma transport – results of the health science research 'study for improving response to severe trauma patients at emergency medical centers'. J Japan Trauma Soc 16(4):319–323
8. Matsumoto S, Funabiki T, Kazamaki T, Orita T, Sekine K, Yamazaki M, Moriya T (2020) Placement accuracy of resuscitative endovascular occlusion balloon into the target zone with external measurement. Trauma Surg Acute Care Open 5(1):e000443
9. Matsumoto S, Hayashida K, Akashi K, Jung K, Sekine K, Funabiki T, Moriya T (2019) Resuscitative Endovascular Balloon Occlusion of the Aorta (REBOA) for Severe Torso Trauma in Japan: a descriptive study. World J Surg 43(7):1700–1707
10. Fukushima F, Moriya T (2021) Objective evaluation study on the shortest time interval from fire department departure to hospital arrival in emergency medical services using a global positioning system — potential for time savings during ambulance running. IATSS Res 45(2):182–189
11. Planning and Development Committee of the Medico-Legal Society of Japan (1997) Report on medico-legal data from the mass-investigation performed by the medico-legal Society of Japan (XIV) autopsy cases of traffic accident in Japan (1990-1994). Jap J Leg Med 52(2):120–126
12. Breen JM, Naess PA, Gjerde H, Gaarder C, Stray-Pedersen A (2018) The significance of preexisting medical conditions, alcohol/drug use and suicidal behavior for drivers in fatal motor vehicle crashes: a retrospective autopsy study. Forensic Sci Med Pathol 14(1):4–17
13. Kurosu A, Kido M, Nagai T, Tokudome S (2006) Epidemiological analysis of natural death while driving in Tokyo (Review). Heart 38(Suppl. 3):61–65
14. Miao Qi, Zhang YL, Miao QF, Yang XA, Zhang F, Yu YG, Li DG (2021) Sudden death from ischemic heart disease while driving: cardiac pathology, clinical characteristics, and countermeasures. Med Sci Monit 27:e929212
15. Hitosugi M, Okubo T (2008) Analysis of sudden onset of disease while driving a taxi. J Jap Council Traffic Sci 8(2):27–32
16. Kagehara B, Shoji T, Takagi T, Okada T, Kashiwade H, Watanabe T, Sudo T, Kajiwara M, Sato Y (1994) Investigation of 4 natural sudden death while driving. J Kyorin Med Soc 25(2):245–248
17. Hitosugi M, Kido M, Kurosu A, Nagai T, Tokudome S (2007) Analysis of sudden death while driving vehicles or bicycles. J Jap Council Traffic Sci 7(1):3–7
18. Scholing M, Saltzherr TP, Fung Kon Jin PHP, Ponsen KJ, Reitsma JB, Lameris JS, Goslings JC (2009) The value of postmortem computed tomography as an alternative for autopsy in trauma victims: a systemic review. Eur Radiol 19(10):2333–2341
19. Ruder TD, Hatch GM, Thali ML, Fischer N (2011) One small scan for radiology, one giant leap for forensic medicine: post-mortem imaging replaces forensic autopsy in a case of traumatic aortic laceration. Legal Med 13(1):41–43
20. Tojo M, Takeda A, Kosou M, Nakamura M, Hitosugi M (2021) Analysis of autopsied sudden cardiac deaths while driving–considering the possibility of survival. J Jap Council Traffic Sci 21(1):40–46

21. Ministry of Land, Infrastructure, Transport and Tourism. Fiscal Year 2021 Business Vehicle Health-Related Accident Countermeasures Council; 2022
22. Charlton JL, Di Stefano M, Dow J, Rapoport MJ, O'Neill D, Odell M, Darzins P, Koppel S (2021) Influence of chronic Illness on crash involvement of motor vehicle drivers, 3rd edn. Monash University Accident Research Centre., Report No. 353
23. All Japan Trucking Association. Health-related accident prevention manual for truck transport operators (revised); 2021
24. Takahashi M (2014) Assisting shift workers through sleep and circadian research. Sleep Biol Rhythms 12(2):85–95
25. Takahashi M (2019) Sociomedical problems of overwork-related deaths and disorders in Japan. J Occup Health 61(4):269–277
26. Japanese Ministry of Health, Labour and Welfare. Outline of the Act on the Arrangement of Related Acts to Promote Work Style Reform. (Act No. 71 of 2018). https://www.mhlw.go.jp/english/policy/employ-labour/labour-standards/dl/201904kizyun.pdf
27. Martin JL, Gadegbeku B, Wu D, Viallon V, Laumon B (2017) Cannabis, alcohol and fatal road accidents. PLoS One 12(11):e0187320
28. Mohamad N, Muhammad M, Haque M, Jawad ZK, Bakar NBA, Ismail A, Simbak N (2016) Traumatic motor vehicle accidents of Malaysia: implications of illicit drugs use. Int Med J 23(2):192–194
29. Funada M (2015) Influence of newly-synthesized psychotropic drug and addictive drugs on driving performance. Jap J Clin Psychopharmacol 18(5):571–576
30. Meola S, Huhtala S, Broseus J, Jendly M, Jalava K, Aalberg L, Esseiva P (2021) Illicit drug profiling in Finland: an exploratory study about end users' perceptions. Forensic Sci Int 324:110848
31. Ezaki J, Taki T, Nakao K (2015) Drug abuse and traffic accidents. IATSS Rev 40(1):35–44
32. World Health Organization. Global status report on road safety; 2018
33. Japan Cabinet Office. White paper on traffic safety in Japan 2022. Status of traffic accidents and current state of traffic safety measures. pp.5, https://www8.cao.go.jp/koutu/taisaku/r04kou_haku/english/pdf/wp2022.pdf
34. Yokoyama A, Omori T, Yokoyama T (2010) Alcohol and aldehyde dehydrogenase polymorphisms and a new strategy for prevention and screening for cancer in the upper aerodigestive tract in East Asians. Keio J Med 59(4):115–130
35. Taylor B, Rehm J (2012) The relation between alcohol consumption and fatal motor vehicle injury: high risk at low alcohol levels. Alcohol Clin Exp Res 36(10):1827–1834
36. Lira MC, Sarda V, Heeren TC, Miller M, Naimi TS (2020) Alcohol policies and motor vehicle crash deaths involving blood alcohol concentrations below 0.08. Am J Prev Med 58(5):622–629
37. Mizoi Y (1990) Forensic medicine and alcohol. Med Philos 9(12):975–983
38. Mizoi Y (1976) Individual differences in the response to alcohol. Jap J Leg Med 30(3):137–168
39. Matsumoto T, Hashitani M, Akane A (2015) Synthetic Cannabinoid and driving impairment. Jap J Clin Toxicol 28(4):333–338
40. Kriikku P, Wilhem L, Schwarz O (2011) New designer drug of abuse: 3, 4-methylenedioxypyrovalerone(MDPV). Findings from apprehended drivers in Finland. Forensic Sci Int 210(1–3):195–200

Open Access This chapter is licensed under the terms of the Creative Commons Attribution-NonCommercial-NoDerivatives 4.0 International License (http://creativecommons.org/licenses/by-nc-nd/4.0/), which permits any noncommercial use, sharing, distribution and reproduction in any medium or format, as long as you give appropriate credit to the original author(s) and the source, provide a link to the Creative Commons license and indicate if you modified the licensed material. You do not have permission under this license to share adapted material derived from this chapter or parts of it.

The images or other third party material in this chapter are included in the chapter's Creative Commons license, unless indicated otherwise in a credit line to the material. If material is not included in the chapter's Creative Commons license and your intended use is not permitted by statutory regulation or exceeds the permitted use, you will need to obtain permission directly from the copyright holder.

Chapter 12
Legal System to Ensure Traffic Safety

Takeyoshi Imai

12.1 Introduction

When it comes to cars, traditional vehicles (driven by humans) are no longer the only ones out there since the self-driving cars (where the controls are entirely or partially controlled by a system called automated driving system) have made their entrances. Nowadays, the use of automated vehicles on public road is attracting attention as the legal system to ensure traffic safety is adapting to new types of vehicles. The fundamental laws at the base of the legal system regarding cars and traffic safety is the Road Transport vehicle Act, the Road Traffic Act, and the Law concerning punishment of acts that cause death or injury by driving a motor vehicle. In this chapter, after an overview of the purpose of these acts and regulatory contents, I will focus on the recent legal developments on measures against drunk driving, dangerous driving, and the preparation toward the use of automated cars.

12.2 The Road Transport Vehicle Law

The Road Transport Vehicle Law is a legislation that regulates vehicle registration, safety standards, inspections, maintenance, maintenance projects, etc. The term "road transport vehicle" as used in the Act is defined in its Article 2 as automobiles, motorized bicycles, and light vehicles.[1] The Act establishes a system for vehicle

---

[1] A light vehicle is defined in paragraph 4 as a vehicle without a motor. Bicycles equipped with a motor are categorized as motorized bicycles but bicycles (including bicycles with electric assistance), cycle rickshaw, horse-drawn carriage, or cart belong to the light vehicle category.

T. Imai (✉)
Hosei University Law School, Tokyo, Japan
e-mail: timai@hosei.ac.jp

registration (Article 4 et seq.) and inspection (Article 58 of the Act) and regulates automobile maintenance and maintenance businesses. The law also defines safety standards and forbids the use of road transport vehicles that don't meet these standards (Article 40 et seq.).

The law was revised in April 2020 and now defines the automatic driving equipment required for automated driving on public roads. Mainly, when used under the conditions (driving environment conditions) set by the Minister of Land, Infrastructure, Transport and Tourism, an automated driving device has the function of replacing all the driver's abilities related to driving, which includes recognition, prediction, decision-making, and operation (Article 41, Paragraph 2 of the Act).

12.3 The Road Traffic Act

12.3.1 Purpose of the Act

The Road Traffic Act aims at preventing road hazards and more broadly at ensuring the safety and fluidity of traffic to prevent any harms that might arise from it.

12.3.2 Legal Definition of Roads

The term "road" is defined in the Road Traffic Act by Article 2, paragraph 1 as follow: "a road as prescribed in Article 2, paragraph (1) of the Road Act (Act No. 180 of 1952), a limited highway as prescribed in Article 2, paragraph (8) of the Road Transportation Act (Act No. 183 of 1951) and any other route used for public traffic." That includes all public roads (roads that can be freely passed by unspecified people and vehicles) such as national expressways, general national roads, prefectural roads, and municipal roads (Article 3 of the Road Act). Private roads (areas built as roads on part of land owned by individuals or corporations) or lands are excluded from the definition of Article 2 unless they are open to unspecified people and vehicles.[2]

12.3.3 Legal Definition of Vehicle

As its goal is ensuring the safety of vehicles' use on public roads, the law defines the vehicles as automobiles, motorized bicycles, light vehicle, and trolleybuses (Article 2, para.1, item 8), and so on.

---

[2] Supreme Court, July 11th, Showa 44(1969)

12.3.4 The System Related to Fines

12.3.4.1 Purpose

The fine system stipulated in the Road Traffic Act concerns relatively minor traffic violations which are resolved by the payment of a fine within a certain period, rather than by a criminal penalty. Since fine is not a criminal but administrative sanction, the offender does not have a criminal record.

12.3.4.2 Overview of the Procedure

The fine system is based on a procedure in which a police officer finds and notifies the violator of the Road Traffic Act of his/her violation and the date and place to make the violator pay the penalty for this violation (Article 126 of the Road Traffic Act). The violator then makes provisional payment or pay the fine after appearing on the date set by the finding notice.[3]

More precisely, when a person commits a violation of the Road Traffic Act, which is subject to the traffic violation notification system, a police officer who witnesses the violation issues a "traffic violation notification form" (so-called "blue kip") and a "provisional fine payment notice" to the violator at the scene of the violation. Fines must be paid in accordance with the notification from the police chief in charge of the area where the violation was committed (Article 127). Provisional payment refers to the payment of an amount equivalent to the penalty for the category including the notified violation, within 7 days after the notification by a police officer or traffic patrol officer (Article 129). The procedure ends with the payment of the fine in the delay fixed by the notification (usually 10 days).

In case of overdue in the payment of the fine, the violator must present himself to the notification center and receive the notification and the fine payment notice. He then must pay the fine within 11 days at the bank or post office, which ends the procedure, and the violator is exempted of a trial (Article 130).

In case of a nonpayment, the case can be trialed under criminal law, if the prosecution decides to indict, which might lead to a guilty verdict and a criminal sanction.

This system which allows a violator of traffic laws to be exempted from criminal trial is limited to minor traffic violations. Therefore, driving without a license, drunk driving, and other offenses that are considered as malicious types of violation are excluded from this procedure (excluded by Article 125, par.2) and are subjected to criminal trial.

---

[3] For the specific procedure, see Chap. 9, art. 125 and following of the Road Traffic Act.

12.4 Measures Against Drunk Driving

12.4.1 Orientations Regarding Drunk Driving

Drunk driving is acknowledged as an extremely dangerous comportment. If the degree of alcohol in the blood passes a certain level, it diminishes the capacity of the driver to drive safely and therefore increases the risks of endangering the life or the body integrity of oneself or other road users. That is why drunk driving is strictly forbidden by the law and subject to criminal sanctions.

The danger posed by drunk driving is something one learns while getting a driver's license, and that is commonly known, yet cases of drunk driving, even though there is a reduction in the overall number, are still an ongoing problem.

In Japan, the number of cases reached a peak of 1276 cases and has been receding until 2021.[4] According to the data released by the National Police Agency,[5] the number of traffic accidents caused by drunk driving in 2022 was 2167, a decrease compared to the previous year (−31 accidents, −1.4% compared to the previous year). Of these, the number of fatal accidents was 120, which also decreased compared to the previous year (−32 accidents, −21.1%). The number of fatal accidents caused by drunk driving has been decreasing significantly since 2002 due to stricter penalties for drunk driving and growing social momentum to eradicate drunk driving, but the rate of decline has slowed since 2008.

The rate for drunk driving fatal accident (number of fatal accidents ÷ number of traffic accidents x 100%) is not only extremely high but also approximately 7.1 times higher than the rate of fatal accidents for non-drunk driving related accidents.[6] Reducing drunk driving therefore remains a major traffic safety issue.

12.4.2 The Following Legal Reform

The government tried to address this issue in different ways, for example by introducing in the Road Traffic Act some criminal offenses and rising the criminal sanctions for drunk driving related offenses.[7]

---

[4] https://www8.cao.go.jp/koutu/taisaku/r04kou_haku/zenbun/genkyo/feature/feature_01_2.html

[5] https://www.npa.go.jp/bureau/traffic/insyu/info.html#:~:text=%E9%A3%B2%E9%85%92%E9%81%8B%E8%BB%A2%E3%81%AB%E3%82%88%E3%82%8B%E4%BA%A4%E9%80%9A%E4%BA%8B%E6%95%85%E3%81%AE%E7%99%BA%E7%94%9F%E7%8A%B6%E6%B3%81%EF%BC%88%E4%BB%A4%E5%92%8C,%EF%BC%8D21.1%EF%BC%85%EF%BC%89%E3%81%97%E3%81%BE%E3%81%97%E3%81%9F%E3%80%82

[6] See the data referred to in note 2.

[7] Takeyoshi Imai "The purpose and remaining tasks for legal measures against drunk driving" Jurist n. 1342, 2007, see p.131.

Also, stricter laws were introduced, such as the 2001 amendment to Article 208-2, Paragraph 1 of the Criminal Code, which criminalized dangerous driving resulting in death or injury. The incrimination was later on moved to the "Act on Punishment of Acts Inflicting Death or Injury on Others by Driving a Motor Vehicle, etc." in 2013. This offense, which involves driving under the influence of alcohol and subsequently causing injury or death, is now subject to more severe punishment. The law applies to drivers who knowingly operate a vehicle while their ability is impaired[8], [9].

These legal incriminations were a reaction to a succession of awful fatal accidents that attracted the attention of society as a whole and led to the establishment of the offense of dangerous driving in the Criminal Code and later on to the establishment of a specific legislation: the "Act on Punishment of Acts Inflicting Death or Injury on Others by Driving a Motor Vehicle, etc." One example of these accidents is the Tomei Expressway Incident. On November 28, 1999, a truck driver drove after drinking alcohol and crashed his vehicle into a regular passenger car in front of him, killing two young girls who were riding in the car. Her parents, who were in the driver and passenger seat, were also seriously injured. At the time, specific offenses for drunk driving in the Criminal Code didn't exist, so the truck driver was charged with professional negligence resulting in death or injury (punished by Article 211 of the Criminal Code)[10] and was sentenced to 4 years of imprisonment. This sentence sparked social debate for not being adequate to the gravity of the accident. The social debate sparked by this case led to the creation of the dangerous driving resulting in the death or injury offense (2001).

A few years later, on August 25, 2006, an employee of Fukuoka City who was driving under the influence of alcohol, collided with a vehicle in front of him. As a

---

[8] Takeyoshi Imai "The establishment of Motor Vehicle Driving Death and Injury Act Punishment Act- Reform of the offense of dangerous driving" Criminal Law Journal, n. 4, 2014, p4.

[9] Criminal offense referring to drunk driving (Overview)
Type of offenses: criminal sanction, administrative penalty
Road Traffic Act:

- Driving under the influence of alcohol (117-2-2)

  1. (alcohol concentration in 1 L of exhaled breath) 0.15 mg or more but less than 0.25 mg: imprisonment for up to 3 years or a fine of up to 500,000 yen, license suspension for 90 days
  2. (alcohol concentration in 1 L of exhaled breath) 0.25 mg or more: imprisonment for up to 3 years or a fine of up to 500,000 yen, license revocation and disqualification period of 2 years

- Drunk driving (117-2): Imprisonment of up to 5 years or a fine of up to one million-yen, license revocation, disqualification period 3 years

  Automobile Driving Death and Injury Activities Punishment Act, art. 2

- Dangerous driving resulting in death or injury.

  – Injury case: up to 15 years imprisonment, license revocation, and disqualification period 5–8 years
  – Fatality case: up to 20 years in prison, license revocation, disqualification period 5–8 years

[10] High Court of Tokyo, January 12th, Heisei 13, Hanrei Jiho 1738, p37.

result, the victim vehicle fell into Hakata Bay, and three children who were on board died (Fukuoka Uminonakamichi Ohashi incident). The Supreme Court confirmed that he had committed a violation of the Road Traffic Act and sentenced him to 20 years in prison, combining both charges of violation of the Road Traffic Act and dangerous driving resulting in the death or injury of other.[11]

Through cases like this applying the newly enacted law, social awareness grew and fatal incidents related to drunk driving generally decreased in number. However, the problem did not completely disappear, and tragic fatal incidents continued to occur. For example, in the prefecture of Hokkaido, there were still many cases involving fatal incidents due to drunk driving.[12] All license holders should know that drinking and driving is extremely dangerous and a form of malicious driving. The recent legislation aimed at reaffirming how vicious and reprehensible drunk driving is. In that sense, lifting the level of the sanction was the right response and its purpose has been fulfilled. Yet, even with a decrease in the number of cases, the publicity regarding malicious and dangerous driving remains and needs to remain an ongoing measure.

12.4.3 The Tasks Ahead

To change the mentality and ultimately reduce drunk driving, social awareness and public attention must be increased and maintained. This is a task that is already undertaken at the national and local level.

The national police agency, by increasing the sanctions against drunk driving, aims at increasing social awareness on the matter.[13]

At the local level, for example, the prefecture of Hokkaido and Fukuoka has shown the resolute stance of eliminating drunk driving, by targeting the drivers, the providers of alcohol, and even the entire society (and potential victims) by the way of ordinance.[14]

---

[11] Supreme Court oct. 31th, Heisei 23, Keishu 65, n.7, 1138

[12] For example, the Otaru dream beach incident (2014, July 13) where 3 women coming home from the beach were killed and another seriously injured because of drunk driving, or the Sunagawa-family incident (2015, June 16) where the perpetrator run at excessive speed through a red light at 60 km/h, entered in collision with the side of a family's car, killing or injuring the 5 passengers inside of it. The driver responsible for the collision had drunk with his friends and then entered in a speed course with them that resulted into violating the traffic signal. The tribunal of Sapporo gave a guilty verdict for the offense of dangerous driving to both parties (the driver responsible for the collision and his friend) and punished them to 23 years of prison. Both defendants appealed the decision, but the appeal was rejected. I will give more details on the case later on.

[13] https://www.npa.go.jp/bureau/traffic/insyu/info.html

[14] See the Fukuoka's ordinance (https://www.pref.fukuoka.lg.jp/uploaded/attachment/50684.pdf) and the Hokkaido's ordinance (https://www.pref.hokkaido.lg.jp/fs/5/4/1/8/3/5/5/_/insyujyourei-honbun.pdf)

These publicity stances might not have any direct effect. Nevertheless, to prevent drunk driving, the constant reaffirmation that drunk driving is a shameful act ignoring society basic rule is undeniably necessary. The social repression of drunk driving should be even more strengthened.

12.5 Measures Against Dangerous Driving

12.5.1 The Establishment of the Dangerous Driving Offense in the Act on Punishment of Acts Inflicting Death or Injury on Others by Driving a Motor Vehicle, etc.

The Act on Punishment of Acts Inflicting Death or Injury on Others by Driving a Motor Vehicle, etc., was published in 2013 and enacted from May 20, 2014. This act is separated from the Criminal Code and specifically targets the offense of voluntary act of dangerous driving resulting in death of injury on others.

12.5.2 Background and Evolution of the Act

On the case previously introduced which led to a sentence in 2001, the driver who caused the death of someone in a car accident was punished under Article 211 of the Criminal Code, i.e., professional negligence resulting in the death or injury of a person (the statutory penalty at that time was up to 5 years of imprisonment or a fine up to 1,000,000 yen[15]). Even in the case where the offense was cumulated with other violation of the road traffic legislation, for example, when a person commits multiple crimes like driving without a license and drunk driving, the maximum sentence is limited. According to Article 47 of the Criminal Code, the sentence for the combined offenses is capped at 1.5 times the sentence for the more severe crime.

Under those circumstances, there were many cases involving a driver who drunk heavily, resulting in many casualties, including children, like the one previously discussed in 4–2 (The Tomei expressway case). That led to the social and legal debate on the appropriateness of the limited criminal punishment, confronting its

---

[15] The amendment in 2002 unified the old concepts of an imprisonment with mandatory labor (Choeki) and one without mandatory labor (Kinko) into a new imprisonment punishment called Koukin, which will be enacted from 2025. The expression imprisonment from here on reflects the up-coming system.

legal basis to the magnitude of the damages. That led to the debate regarding lifting the sentence and creating a more specific offense.

As a result, in 2001, Article 208-2 punishing dangerous driving was added to the Criminal Code.[16] Later, in 2007, the Criminal Code was amended, introducing a second paragraph to Article 211, para. 2 (negligent driving) lifting the previous professional negligence sentence to up to 7 years of imprisonment and the fine up to 1,000,000 yen.

Later on, in 2013, both newly introduced articles were taken out of the Criminal Code, and moved into the new act, more precisely in its Article 2 punishing dangerous driving and Article 5 punishing negligent driving.[17] The new Act was enacted in 2014 (without changes to the statutory penalties).

Furthermore, in 2020, the Act on Punishment of Death and Injury by Motor Vehicle Driving (Article 2) was amended, and the types of acts subject to punishment (so-called reckless driving as a dangerous act) were revised. Two new items were introduced to Article 2 as items (5) and (6) (the statutory penalties remain unchanged), and what used to be items (5) and (6) were downgraded to items (7) and (8), respectively, leading to the following enumeration of acts subjects to punishment.

12.5.3 The Types of Dangerous Driving

Article 2 of the Act[18]

---

[16] The legal punishment has been hardened and is now up to 15 years of imprisonment in case of injury and from 1 up to 20 years in case of death.

[17] The expression of vehicle has been suppressed because, in the act punishing the acts of vehicle driving resulting in the death or injury, etc., adding the term vehicle to driving is not important and does not affect the content of the disposition.

[18] More precisely the disposition gives an enumeration of what is considered as "dangerous driving."

5–2 Acts falling under the punishment provision.

The dangerous driving described in Article 2 (of the same law) targets the following acts:

(i) Act of driving a motor vehicle under the influence of alcohol or drugs making it difficult for the person to drive safely.
(ii) Act of driving at such high speed that it is exceedingly difficult for the person to control the motor vehicle.
(iii) Act of driving when the person lacks the skills to control the motor vehicle.
(iv) Act of, while driving a motor vehicle, cutting in directly in front of another running motor vehicle or otherwise approaching in close proximity to a passing person or vehicle, with the intent to obstruct the passage of another person or vehicle, at a speed that can cause serious danger to traffic.
(v) Act of stopping in front of or approaching in close proximity of a driving motor vehicle with the intent to obstruct the passage of a vehicle.
(vi) Act of stopping in front of or approaching in close proximity of a driving motor vehicle to make it stop or slow down in the expressway with the intent to obstruct the passage of a vehicle.

No. 1: Alcohol or drug influence type
No. 2: Difficult to control high speed type
No. 3: Unskilled type
No. 4: Type of approach such as intent to obstruct traffic, last-minute intrusion, etc.
No. 5: Act based on the intent to obstruct traffic, approach type such as stopping (newly established in 2020)
No. 6: Act based on the intent to obstruct traffic, expressway type (newly established in 2020)
No. 7: Special disregard for red lights
No. 8: Prohibited road progression type

12.5.4 Background of New Categories (Article 2, (5) and (6)) Related to Tailgating Driving

The incident that brought up for the 2020 revision of the law was the so-called "Tomei Expressway reckless driving accident" that occurred on June 5, 2017. It happened in the overtaking lane of the Tomei Expressway, when the defendant's car forcibly stopped the victim's car and assaulted the victim. A truck then arrived from behind and crashed into the victim's car, which resulted in the death of a married couple and injuries to their two children. Beside the trunk driver, the question was on how to punish the defendant, for forcibly stopping the car in the expressway which is quite hazardous. However, per the principle of legality, in order to apply the dangerous driving offense, one of the acts described in the Act on Punishment of Acts that Punish Death or Injury to Motor Vehicles (Article 2 of the same act) must be proven. Here, the act considered was the driving at dangerous speed targeted in Article 2(4): "the act of driving a motor vehicle at a speed that creates a traffic hazard." The main legal issue was to know if, even if a death or injury accident occurs due to a speed of 0 km/h (when the car is not running (stopped)), it can still be considered as the offense of dangerous driving due to a speed creating a traffic hazard. More simply put, the point requiring to be clarified was whether no-speed can be considered as a speed under this article. The question was debated but

---

(vii) Act of driving a motor vehicle deliberately ignoring a red signal light or its equivalent at a speed that can cause serious danger to traffic.
(viii) An act of driving a motor vehicle through a passage-prohibited road (meaning a road or a part thereof on which motor vehicles are prohibited from passing through by road signs or road markings, or pursuant to the provisions of laws and regulations, where Cabinet Order specifies that the passage through such road or a part thereof will cause traffic danger to persons or vehicles) at a speed that can cause serious danger to traffic.

prosecuting for causing death or injury caused by dangerous speed in a case of no-speed at all reveled itself to be quite difficult.

The court of first instance concluded that the defendant's series of actions in which he repeatedly blocked the victim's vehicle's path four times on the expressway and stopped the victim's vehicle caused a rear-end collision with the victim's vehicle (by the trunk that was following him). He was found guilty of dangerous driving resulting in death and injury, as there was a causal relationship with the victim's death. The court of second instance overturned the judgment of the first instance, finding that the proceedings of the first instance had violated laws and regulations, and the case was remanded to the court of first instance. After the case was remanded, the first instance judge recognized the causal relationship between obstructive driving and the accident and found that the crime of dangerous driving resulted in death or injury, but the defense appealed. Therefore, the case has not yet (at the time of the redaction, i.e., Nov 2023) been concluded. Furthermore, as mentioned earlier, this case occurred before the 2001 amendment to Article 2, Items (5) and (6) of the same Act. Therefore, the new offense of traffic obstruction cannot be applied retroactively to this incident.

12.5.5 How to Tackle Cases Where Dangerous Driving Is Excluded

Dangerous driving resulting in death or injury is committed, as per the name of the offense, when dangerous driving results in injury or death to the victim. Therefore, if the violation does not result in death or injury to the victim, then the offense is only punishable under the Road Traffic Act. In other words, there is not a special type of dangerous driving punished as such distinct from the Road Traffic Act.

For example, if the driver was engaged in reckless driving but no one was killed or injured, the offense of obstructive driving, which was newly established in the 2020 revised Road Traffic Act, could be applied. In other words, item (11) of Article 117-2-2, of the same law punishes a person who drives for the intent of obstructing the passage of other vehicles with imprisonment for up to 3 years or a fine of up to 500,000 yen.

In addition, items (2) and (6) of Article 117 establish that a person who causes a significant traffic hazard by violating article 11 of the same law (distracted driving to obstruct traffic such as a procession) shall be sentenced to up to 5 years of imprisonment or a fine up to one million yen.

12.6 Amendments of the Law Related to Automated Driving

12.6.1 The Amendment of the Road Traffic Act

The 2019 amendment of the Road Traffic Act[19] recognizes for the first time in the world the use of so-called level 3 of automated driving.[20] However, the passenger is still considered as the all-time driver and is not relieved of his duty to monitor the environment. In that aspect, the qualification of level 3 for this type of driving is subject to discussion, even on the international level. Indeed, it is not debated that under level 2, the passenger seated on the driver seat is considered a driver and thus subject to the traditional driver's obligations under the Road Traffic Act. However, the fact that those obligations remain the same for the person in the driver seat, even under the newly admitted level 3, is more questionable.[21]

That depends on the understanding one has regarding to level 3. If the understanding is that the level 3, although it leaves the possibility of a take-over request in case of emergency where the passenger is called to take over as a driver, is otherwise equivalent to level 4 while the automated driving mode is activated. In other terms, outside of these situations requiring a TOR, the vehicle is completely controlled by the system, which means the actual implementation of automated driving and not a form of assisted driving. The technology for level 4, and even more so with level 5,[22] allows the passenger to freely use his time while the automated driving mode is activated. Based on that understanding, not lifting the traditional obligations for the driver in level 3 when the automated driving mode is activated (i.e., basically similar to level 4) seems contradictory. On the other hand, from a policy perspective, especially with the utmost attention toward traffic's safety, this choice made by the amendment in 2019 is understandable. In order to allow self-driving technology use on public roads, both improving the convenience of the passenger (mainly the free use of his time) and maintaining or raising the level of traffic's safety are necessary. In that perspective, considering that the 2019 amendment took both elements into consideration, with an emphasis on the second, in order to adopt a practical policy is not too far-fetched.

---

[19] Legislation n.20, 2019 (first year of the Reiwa period).

[20] SAE J3016, Apr 2021 (Surface Vehicle Recommended Practice), p 17 (3.20.).

[21] Takeyoshi Imai "Evolutions towards the implementation of self-driving cars" Horistsu no Hiroba, Feb. 2022, pp. 44–45.

[22] Level 4 implements the automated driving technology only inside the ODD (operational design domain), meaning the prerequisite conditions regarding the environment that needs to be met to use the automated driving mode. In level 5, the ODD is not limited and cover all areas and environment.

12.6.2 The 2022 Revision of the Road Traffic Act

12.6.2.1 General Observations

The 2022 revision of the Road Traffic Act[23] went further along in the automated driving legalization, with the introduction of a system allowing the use of level 4. In this new system, the "operation" part of the driving activity (which includes as previously mentioned recognition, prediction, decision-making, and operation) is described as a specific automated operation. Its implementation by the automated operation equipment is carried out by the specific automated operation operator, under and with the help of the specific automated operation chief and the on-site measures worker. A monitoring system was also introduced. The specific automatic operation is conceptually included in the broader category of "driving," which should require the identification of the driver. The legal definition of the driver is less certain, and there seems to be, at least until now, very little academic interest in Japan.

12.6.3 Definition of "Specific Automated Operation"

Specific automated operation is defined in the Road Traffic Act as the operation where the equipment controls the equipped vehicle under the operation's conditions of use on the road[24] (except for the times when a person controls the vehicle to adapt to a particular situation of the road, whether it is due to the traffic or the vehicle itself). The equipment in this definition is limited to the vehicle to the one that is able to execute a so-called minimal risk maneuver, which means it is able to stop directly and automatically, by bringing the car to a stop in a safe manner, if the vehicle equipped falls into the category "poorly maintained" described by Article 62, or if the conditions of uses for automated operation are no longer met.

Specific automated driving is based on an authorization delivered[25] by the prefectural public safety commission[26] after review of the specific automatic driving operation plan.[27] During the reviewing process, the commission is required to hear opinions from the heads of localities (city and villages) included in the automated service routes, as to whether such transportation is deemed to contribute to

---

[23] Legislation n.31, April 27th, 2022.
[24] Road Traffic Act, art, 2, par. 17 (2).
[25] Road Traffic Act, art. 75-12, par.1.
[26] Road Traffic Act, art. 75-13, par. 1.
[27] Road Traffic Act art. 75-2, al. 2 (2).

improving the convenience of or welfare of local residents.[28] If a specific automated operation operator or one of his employees violates an order or a legal disposition, the prefectural public safety commission shall instruct the specified automated operation operator, revoke his permit, or issue an order to suspend the validity of the permit.[29] In addition, if there is an urgent need to prevent hazards on the road, the police chief may temporarily suspend the validity of the permit.[30] Specific automated operation is not limited to the transport of persons and can now also be used for cargo transportation.[31]

In that sense, specific automated (driving) operation is not classified as driving but as an operation, which allows for the persons related to its implementations to not be submitted to the driver's obligations of the Road Traffic Act.

12.6.3.1 Specific Automated Operation Operator, Specific Automated Operation Chief, On-Site Measures Operator

The revision of the Road Traffic Act introduced three new agents, called specific automated operation operator, specific automated operation chief, and on-site measures operator. Hereafter is a brief check on the functions of each new entity.

1. Specific automated operation chief
   The specific automated operation chief can be either in the car or on a remote location.[32] During the automated operation, he has the duty to monitor the state of the automated operation equipment and to take any urgent measures to end the operation if the equipment is not functioning normally.[33]
2. The on-site measures operator
   In prevision of any incident involving the specific automated operation, the specific automated operation operator designates an on-site measures operator.[34] When a traffic incident occurs, the specific automated operation operator must

---

[28] Road Traffic Act, art. 75-13, par. 2 (2).

[29] Road Traffic Act art. 73-26, art 75-27

[30] Road Traffic Act, art. 75-28

[31] Road Traffic Act, art. 75-12, par. 2 item 2. See below for details on how cargo transportation should be carried out using specified automated operation.

[32] Road Traffic Act. art. 75-20 par. 1. (1) of the same article gives some specifications to take into account in case of remote driving.

[33] Road Traffic Act. Art. 75-21, par 1.

[34] Road Traffic Act, art. 75-19, par. 3

notify the fire or police department and dispatch the on-site measures operator.[35] The dispatched on-site measures operator takes all necessary measures to prevent any road hazards.[36] If during the traffic incident, the specific automated operation chief was in the vehicle, he must report to the police department of the number of injured persons and aid injured persons. [37],[38]

3. The specific automated operation operator

The specific automated operation chief and the on-site measures operator could be considered as the executioners of the specific automated operation operator. However, even if the operation operator uses them to control the vehicle, it is difficult to consider him as the driver. The presence or identification of the driver leads us to the underlying question of knowing if this "operation" can be considered as self-driving or something else.

The 2022 amendment of the Road Traffic Act choses to bypass the issue of the mandatory driver when it comes to driving, by classifying the newly introduced system as "operation" instead of "driving." Based on this classification, which favors the second option (something else), neither the automated operation operator nor the chief is considered the driver. As a result, both are excluded from offenses related to dangerous and negligent driving. However, what if the specific automated operation chief who is responsible for monitoring the operations is not in a normal state allowing him to adequately control the situation or he is not qualified to perform the task (under qualification)? If an incident resulting in the injury or death occurs under these circumstances, can we consider this exclusion from the current law which punishes the "driver" for dangerous "driving," a good thing? It is definitely one of the remaining legal tasks ahead.

This distinction between "driving" (which requires a driver) and "operation" (which does not) is possible under Japan national law. However, in order to avoid violating its mother law, that is, the international Geneva Convention on Road Traffic (1949), it would be preferable if there were a disposition in the convention allowing members states to adopt these types of distinctions.[39] The specific automated operation operator introduced by the 2022 Road Traffic Act amendment

---

[35] Road Traffic Act. Art. 75-23, par. 1

[36] Road Traffic Act. Art. 75-23, par. 2 and 5

[37] Road Traffic Act. Art. 75-23, par 3 and 5

[38] The other duties of the specific automated operation operator are listed in the Road Traffic Act art. 75-20, par, 2 and art, 75-22

[39] In the Vienna Convention on Road Traffic (1968), the possibility for member states to introduce in their national laws a disposition to allow operation without a driver present has been introduced (art 34 bis, para. 1 pose that the requirement for a driver is met if the self-driving car complies with domestic technical regulation and domestic driving regulations related laws and regulations). On the other hand, Japan is a member state of the Geneva Convention, which does not have a disposition like art 34 bis. Therefore, introducing a concept like the technical supervisor introduced in Germany based on Article 34 bis of the Vienna Convention, without such a disposition seems to be in violation with the Geneva Convention. However, since there is a resolution between the members of the Geneva Convention to interpret it as if it was allowing such dispositions, there is a margin of interpretation to exclude the risk of a violation of customary international laws.

imposes to the specific automated operation chief working under him an obligation to remotely monitor the operation executed by the vehicle. Also, in the case where the automated driving system (ADS) has reached its limits or the vehicle is malfunctioning, the equipment should automatically and directly stop the vehicle in a safe manner.[40] In other words, in the occurrence of the ADS reaching its limits, the vehicle itself chooses and executes the so-called minimal risk maneuver (MRM). If the ADS is the one who executes the MRM, it means that the specific automated operation chief does not actually intervene in this MRM, allowing the categorization for this vehicle equipped with the ADS of level 4 self-driving vehicle. On the other hand, if you focus on the executing part of the operation, that is also possible to put in the same category the system where the specific automated operation operator authorizes from a remote place the "driving" operation beforehand, and the ADS executes it. To properly categorize Level 4 self-driving technology, we first need to define what a remote driver is. This is a critical task for the future.

---

[40] Id. Road Traffic Act, art. 2, para.17-2

Open Access This chapter is licensed under the terms of the Creative Commons Attribution-NonCommercial-NoDerivatives 4.0 International License (http://creativecommons.org/licenses/by-nc-nd/4.0/), which permits any noncommercial use, sharing, distribution and reproduction in any medium or format, as long as you give appropriate credit to the original author(s) and the source, provide a link to the Creative Commons license and indicate if you modified the licensed material. You do not have permission under this license to share adapted material derived from this chapter or parts of it.

The images or other third party material in this chapter are included in the chapter's Creative Commons license, unless indicated otherwise in a credit line to the material. If material is not included in the chapter's Creative Commons license and your intended use is not permitted by statutory regulation or exceeds the permitted use, you will need to obtain permission directly from the copyright holder.

Chapter 13
Sustainable Growth: An Economic Perspective

Kazuhiro Ohta, Mariko Futamura, and Akihiro Nakamura

13.1 Two New Technological Innovations

While the economic perspective of improving society by effectively utilizing scarce resources remains unchanged, two major foundational technological innovations have emerged in the transportation society over the past decade. These two technological innovations have not yet prevailed in society, but they are set to fundamentally change the nature of transportation society in future.

First, the spread of information communications technology (ICT) can be mentioned. The global pandemic of the novel coronavirus has rapidly popularized various online meetings. In the field of education, online classes were forcibly introduced, and as a result, the need for commuting disappeared. After the end of the pandemic, the indispensability of face-to-face communication is being reconsidered, but the substitution of transportation by the advancement of information and communication technology is expected to progress in business activities. If the mainstream of people's mobility needs shifts from business trips to tourism trips, the nature of transportation society must also change.

The second technological innovation is the advancement of the social implementation of autonomous driving technology. The realization of fully autonomous driving is still not in sight, but this trend of technological innovation is inevitable. The technology of fully autonomous driving changes trade conditions, and therefore changes the division of labor within a country. If it is a contiguous country, it will have a significant impact on the international division of labor. At present, it is

---

K. Ohta (✉)
School of Commerce, Senshu University, Tokyo, Japan

M. Futamura
Department of Economics, Tokyo Woman's Christian University, Tokyo, Japan

A. Nakamura
Faculty of Economics, Chuo University, Tokyo, Japan

© The Author(s) 2026
Pioneering the Future for Traffic and Safety Sciences,
https://doi.org/10.1007/978-981-96-0676-4_13

impossible to foresee the impact of fully autonomous driving, but it is meaningful to foresee the future social changes.

13.1.1 Traditional Economic Perspective

Economics assumes that both consumers and suppliers act to maximize their own net benefits (benefits-costs), and such behavior is considered rational. For example, it is assumed that a transportation service provider will supply an additional unit of transportation service if the incremental revenue gained when supplying an additional unit of transportation service exceeds the incremental costs incurred by doing so. On the other hand, it is assumed that consumers will also purchase an additional unit of transportation service if the increase in utility gained from consuming an additional unit of transportation service exceeds the increase in cost they pay for it.

Thus, in traditional economics, it is thought that market transactions occur only when benefits exceed costs, or in other words, when there are positive net benefits to the society. Therefore, it is believed that a fully functioning market (perfect competitive market) would lead to an efficient allocation of resources as a result of people freely trading in the market.

13.1.2 Market Failure

However, a market is not so perfect. There are many instances where there is some dysfunction in the market that does not lead to efficient resource allocation. This is called market failure, and because it is a failure, policies are needed to correct it. There are cases where the market does not bring about appropriate resource allocation, which is defined as a narrow sense of market failure. In addition, even if a market functions properly, what the market brings is efficient resource allocation only, and even if society desires a distribution of wealth other than that, markets do not have the function to achieve it. Therefore, there is a broad sense of market failure in the sense that the market cannot achieve fair distribution, even if efficient resource allocation has been achieved.

Among the narrow market failures, externality is a case where benefits or costs occur to third parties other than the consumers and suppliers participating in the transaction without going through the market. Typical social problems due to externality include environmental problems and congestion problems, which will be discussed later. Traffic accidents are another case where public intervention is necessary due to externality. Also, natural monopoly, which will be explained later, is often observed as a market failure in the transportation sector and is the basis for policy intervention or regulatory policy.

In addition, the information asymmetry, where service users cannot judge the safety of the transportation service in advance, is also one of the market failures and is one of the grounds for regulating the safety aspect of the transportation sector. In general, consumers who do not pay fees can be excluded in service transactions, but for example, it is difficult to collect fees each time on city roads. Therefore, cost-benefit analysis is necessary to evaluate the social benefits compared to the construction cost in order to decide whether to build the infrastructure as a matter of policy. This is a perspective of the supply of public goods, one of the market failures. Thus, the transportation market is a field where market failures occur relatively frequently, and it is a market with many regulations, which is a form of policy intervention.

In the transportation sector, measures are often taken to correct a broad sense of market failure. Market transactions are the result of competition, so there are losers. If society decides that vulnerable populations should be guaranteed mobility, public responses will be taken from the perspective of fairness.

13.1.3 Various Regulations on Transportation Services

As mentioned in the previous section, generally, transportation services are said to be prone to public intervention, or regulation. Regulations are broadly classified into "social regulations" and "economic regulations." While social regulations are qualitative regulations such as safety regulations and environmental regulations, economic regulations are quantitative regulations that affect supply and demand through the market. Economic regulations are categorized as price regulations and entry regulations, and transportation services in particular have been subject to supply and demand adjustment through these regulations, which have publicly restrained competition. However, in recent years, deregulation has been promoted worldwide, and transportation markets are now aiming to become more efficient through competition.

The transportation market has often been subject to economic regulation due to its characteristics of natural monopoly. Not only in the transportation market, but also in other industries with large fixed costs, economies of scale are likely to exist, and if companies compete freely, it is expected that they will engage in extremely intense "destructive competition" in terms of price reductions and service competition. Then, after the losing competitors have exited the market and a natural monopoly has been achieved, prices can be raised and monopoly profits can be enjoyed. As a result, consumers face monopoly prices and services are undersupplied. Also, the "sunk costs" that cannot be recovered from the investments of competitors who have exited the market become a loss for society.

The transportation market often has the above characteristics, and due to the high necessity of transportation services, the impact of any problems is expected to be large. Therefore, companies have been allowed to monopolize or oligopolize in exchange for price regulation. Price regulation has been based on the concept of "full cost principle," in which necessary revenue is defined as the cost incurred under efficient management plus an appropriate profit margin.

However, the full cost principle does not provide incentives for cost reduction to operators, but rather leads to inefficiencies in management. In particular, the application of fair rate of return regulation has led to overinvestment and increased costs. On the other hand, public intervention in the transportation market has changed due to the theory of contestable markets proposed by Baumol and others since the mid-1970s.

A contestable market is a market where "entry and exit are free and there are virtually no sunk costs, so potential competition weakens the monopoly power of incumbent operators".[1] And if it is perfectly contestable, existing operators are said to behave efficiently for the following reasons.

If the incumbent company is enjoying excess profits in a monopoly or oligopoly market, a new company can enter the market at a lower price than the incumbent company, take all customers, and enjoy profits. Then, if the incumbent company counters by offering the same or lower price, the new entrant will lose its profits and exit. The theory states that even if an incumbent company is in a monopolistic market, if new entrants are free to enter and exit the market and there are no sunk costs, the threat of potential new entrants will cause incumbents to set efficient prices.

13.1.4 Progress of Deregulation

The theory of contestable markets was applied to U.S. domestic air services in 1978, and this led to worldwide progress in deregulating the transportation sector. The bus and airline markets in the United Kingdom and the airline, cab, and truck markets in Japan have also been deregulated.

The aviation industry in Japan shifted from a policy of suppressing competition to a competition policy in the mid-1990s, with deregulation of entry and price controls being promoted. As a result, new operators have entered the market one after another, and existing airlines and other carriers have diversified their prices and services. However, in the case of Japan's aviation industry, free entry, which is the premise of the contestable market theory, has not been established due to airport congestion, such as at Haneda Airport. In such cases, the role of public entities in the use of airports by airlines remains significant.

---

[1] Quoted from Button [1].

13.1.5 Perspective of Behavioral Economics

In contrast to the traditional economics discussed so far, the perspective of behavioral economics has been attracting attention recently. Market failure refers to a situation where the transaction outcome is not socially desirable in terms of the results of rational individuals acting freely. Therefore, public intervention is discussed to complement the functioning of markets.

On the other hand, behavioral economics begins with a focus on the fact that there are behaviors that systematically deviate from the rational individual behavior assumed by traditional economics. With the typology of cases where people systematically do not take rational actions becoming somewhat established, a system of behavioral economics is being formed.

Insights from behavioral economics, rather, have a perspective of understanding cases where people do not act rationally, raising awareness of them and encouraging more rational behavior, and a perspective of using the behavioral patterns that people systematically take to guide them well using nudges. The definition of a nudge is somewhat vague, but it often refers to cases where behavior change is encouraged using psychological techniques other than methods using monetary incentives such as fines.

In fact, not only in traffic behavior, but also when trying to encourage behavior change with monetary incentives such as fines, a wrong interpretation is that it is okay to do undesirable things because a fine is paid, and it may even lead to a reversal of the original intention. In such cases, behavioral economic methods are often considered.

13.2 Cost-Benefit Analysis

In the case of goods and services supplied by the private sector, the sector makes a decision on whether or not to supply them by considering the income and costs associated with the supply. However, in the case of transportation infrastructure, there are many cases where the benefits spill over to those other than the users, or where it is difficult to collect fees from the users. In such cases, cost-benefit analysis (CBA) is conducted.

13.2.1 Purpose and Basic Concept of Cost-Benefit Analysis

Cost-benefit analysis (CBA) serves as a tool for evaluating the economic impacts of infrastructure development, including roads and railways, and for guiding decisions concerning project implementation [2]. A federal waterway project by the Tennessee

Valley Authority (TVA), which was begun during the 1930s as part of the New Deal policy in the United States, was the first systematic evaluation using CBA. The purpose is to confirm whether a sufficient development effect would be realized when investing large amounts of tax money required to construct dams and other massive project.

Thus, cost-benefit analysis is applied not only to transportation facilities but also widely to the project evaluation of public investment. The basic idea is to calculate and compare in terms of monetary units the costs required for a project and the benefits derived from it. In the case of transportation facilities, the monetary costs include the development costs (construction costs) and ongoing maintenance and management costs, and other costs, which are paid by the public entities through taxes. There are also cases where negative effects to society resulting from the project, such as environmental degradation, are included in the calculation as additional costs or negative benefits.

13.2.2 Evaluation Criteria in Cost-Benefit Analysis

Because of limitation of project budgets (financial resources), CBA is applied not to determine whether projects should be carried out, but rather to make a relative comparison for selecting between multiple alternatives. There are three main criteria for project evaluation.

13.2.2.1 Net Present Value Method

This is a method of assigning priority based on the amount of net present value calculated by the following equation:

$$NPV = B - C - K = \sum_{T}^{t=0} \frac{B_t}{(1+i)^t} - \sum_{T}^{t=0} \frac{C_t}{(1+i)^t} - K$$

Here, *NPV* is the net present value, *B* is the total benefits present valued at base year 0, *C* is the total maintenance costs present valued at base year 0, *K* is the construction costs present valued at base year 0, t is the tear, *T* is the evaluation term (service life or durability in years), B_t (C_t) is the t year's benefits (maintenance costs), and i is the social discount rate.

13.2.2.2 Cost-Benefit Ratio Method

This is a method of assigning priority by calculating the ratio of cost to benefit. In practice, two ratios can be used: $(B - C)/K$, which is the net benefits per units of investment(construction costs), and $B/(C + K)$, which is the ratio of gross benefits to total costs. As both ratios represent cost-benefit ratios, it is necessary to choose in advance which ratio is used for the project appraisal.

13.2.2.3 Internal Rate of Return Method

Internal rate of return indicates how much the rate of return from an investment increases; the larger the internal rate of return, the better the project. The internal rate of return r is the solution to the following equation:

$$\sum_{t=0}^{T} \frac{B_t}{(1+r)^t} - \sum_{t=0}^{T} \frac{C_t}{(1+r)^t} = K$$

The internal rate of return is the criterion most consistent with economic theory. However, if there is a significant cost at the end of the evaluation period (for example, the cost of decommissioning a nuclear power plant), the internal rate of return may sometimes have multiple possible solutions. However, in the case of transportation improvement projects, this possibility is small, so this method is recommended for the project evaluation.

13.2.3 *Benefit Items*

In the case of transportation facilities, the main benefits are time-saving benefits, cost reduction benefits (for example, fuel savings in the case of road improvement), safety improvement benefits, and comfort improvement benefits. Due to the nature of transportation facility improvement, time-saving benefits make up a large part of the total benefits. When calculating benefits, it is important to set the evaluation value per unit of each benefit as it can influence the evaluation results.

13.2.3.1 Time-Saving Benefits

For passengers, time saving for movement makes them use the saved time for other activities. As for logistics, time saving for transportation benefits reduced interest payment and insurance fees for transportation.

Since a large part of the effect of transportation facility improvement is time-saving, there has been accumulated research [3]. The opportunity cost of time (income obtained when the saved time is redirected to other productive actions) tends to increase with economic growth, so profound considerations and studies are needed in setting the value of time in developing countries.

Also, if autonomous driving is realized, productive activities can be carried out even while moving in a car. Whether or not to consider the saving of travel time in a fully autonomous driving car as worthless is a matter of debate. It is possible to argue that it is not worthless, as time saved has been accounted for as a benefit in movement on airplane trip or bullet train. However, this traditional method of time value evaluation itself has issues. The improvement in the value evaluation of saved time is required according to the era, technology, and the development stage of the country.

13.2.3.2 Cost Reduction Benefits

If the flow of traffic becomes smoother, the monetary cost required to move a certain distance (mainly fuel costs) will be reduced. Directly, the reduction in the use of fossil fuels will be accounted for as a benefit. If the electrification of cars progresses, the reduction in electricity usage will be accounted for.

As a method, cost-benefit analysis is grounded in technologies and consitions available at the time of evaluation, thereby reflecting the perspectives of the present generation rather than those of future generations. However, this does not mean that future technological changes can be ignored. If the transition from fossil fuel vehicles to electric vehicles progresses, the evaluation unit value will change. Also, if the fuel efficiency of vehicles (whether fossil fuel or electricity) improves, the benefits of reducing travel costs due to transportation infrastructure development will decrease.

In any case, it is necessary to clearly assume the evaluation unit.

13.2.3.3 Safety Improvement Benefits

If the number of traffic accidents or the degree of damage decreases due to transportation infrastructure development or traffic safety measures, these facility development or measures improve safety. Traditionally, human losses due to fatal accidents have been accounted as safety improvement benefits by considering them as a loss of productivity. The same approach has been adopted to measure the loss of productivity (lost profits) for injury accidents.

It can be ethically controversial to evaluate humans as if they are tools for conducting economic activities and increasing GDP. Therefore, a new approach is being sought for the evaluation of human life.

13.2.3.4 Comfort Improvement Benefits

If humans are seen as tools of production, the comfort of movement is not focused on. However, as the economy grows, people will desire a higher quality of life. Here, as economic development progresses, comfort is emphasized and tends to be evaluated as a benefit.

Although there are many proposals for measuring comfort, none of them are necessarily established. However, the approach of objectively measuring comfort benefits from people's choice behavior is becoming mainstream.

13.2.4 Foreign Exchange Rate Issues

In cases where funds are raised domestically, the evaluation methods described above can be applied regardless of whether funding will be through taxes or bonds. On the other hand, in cases where funds are borrowed from abroad, it is necessary to consider future foreign exchange rate fluctuations. In particular, in developing countries, it is essential to confirm the ability to repay foreign debt in project evaluation [4, 5].

Also, if the energy for movement, i.e., fossil fuels or electricity, depends on procurement from abroad, this cost will be influenced by foreign exchange rate. In addition, it is affected by the international energy prices.

How much of these points to include in the cost-benefit analysis, or whether to reflect them in the project evaluation outside the cost-benefit analysis, needs to be flexibly responded to according to each situation (country or region).

13.3 Traffic Congestion and the Policy (Economic Measures)

13.3.1 What Is Congestion?

Congestion is the phenomenon of the result of a concentration of relatively large demand in consumption to the capacity of facilities on equipment, and causes many monetary and nonmonetary costs that do not arise in cases of no congestion. In the transportation area, these include problems such as crowding on trains, congestion on road networks, congestion at airports related to aircraft takeoff and landing slots, and congestion of ship entry at ports. In any case, it is assumed that the levels of infrastructure supply cannot be changed in the short term. That is, collective consumption is possible up to a certain level, but exceeding that level, external diseconomies occur, hindering the original use. Among various congestion-related problems, this chapter deals with the problem of road congestion that many cities around the world suffer.

13.3.2 Economic Understanding of Congestion

Roads are goods that can be collectively consumed up to a certain level, and users can drive smoothly. However, when a road becomes congested, the driving speed on the road decreases, and the time and costs such as gasoline expenses to reach each car's destination increase.

In Fig. 13.1, a graph illustrating economic concepts with three intersecting lines labeled as the Marginal Social Cost Curve, Marginal Private Cost Curve, and Demand Curve. The x-axis is labeled "O" to "F" with traffic flow and the y-axis is labeled "O" to "C" with costs. Points E, A, B, and G are marked, with a shaded area between the curves. Road capacity F_0 is the upper limit of the traffic volume that can be consumed equally, and when automobiles in excess of this limit enter into the road, the marginal cost begins to increase. That is, with the inflow of additional traffic, each driver faces a decrease in driving speed and an increase in required time, and gradually recognize the increase in private marginal cost. The graph demonstrates the relationship between social and private costs in relation to demand. All drivers experienced the increase cost resulting from the inflow of additional traffic in a similar way, so the marginal social cost curve diverges. Thus, the realized traffic volume is F_1, deadweight loss exist, and a situation where social surplus loss occurs.

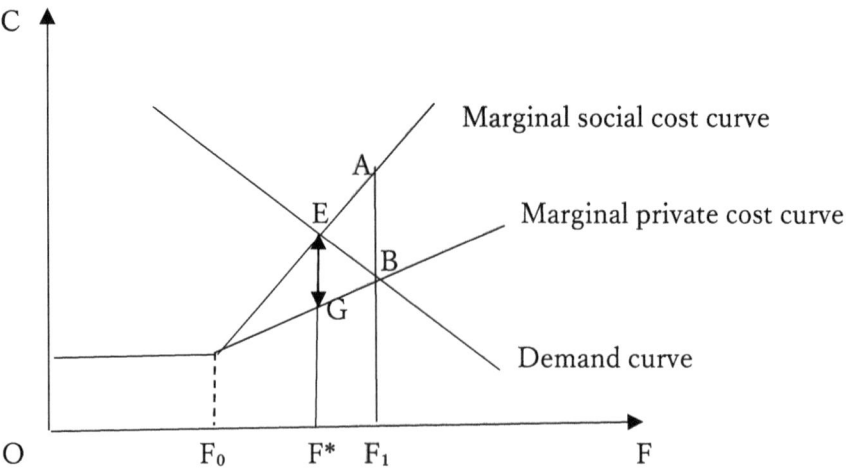

Fig. 13.1 An illustration of vehicle congestion

13 Sustainable Growth: An Economic Perspective   245

This congestion is caused by external diseconomies and can be internalized by imposing a tax rate t equivalent to EG in the figure according to the Pigovian tax, and the optimal congestion level F* can be achieved. Note that the social surplus is maximized at this F*.

13.3.3 Introduction of Congestion Charging

Road congestion is a major problem in many cities around the world, and the introduction of congestion charging is often discussed as a solution. In fact, many cities such as Singapore, Oslo, and London have introduced congestion charging to downtown areas. And the introduction to London represents the first case implemented in a major of a developed country.

Beginning with the 1964 Smeed Report, there have been many studies and discussions on introducing a road congestion in London, but establishment of Greater London Authority in 1997 and the 2000 election of a mayor who supported the idea laid the groundwork for actual implementation. In 2003, the London Congestion Charging Scheme was introduced with the aim of improving traffic flow [6]. Under this scheme, the charge was implemented by the Transport for London, the charge for inflow into the designated area was 5 pounds per day (from 7 am to 6 pm on weekdays).The charges were incrementally raised, as of October 2023, 15 pounds per day (from 7 am to 6 pm on weekdays, Saturdays, Sundays, and additionally from 12 pm to 6 pm on bank holidays). It should be noted that the use of the revenue from the charge is limited to public transportation. The scheme is now in its 20th year, and the London Transport Authority's website shows its effects: "An 18% reduction in inflow into the zone during weekday charging hours, a 30% reduction in congestion, a 33% increase in bus usage in central London, and a shift of 10% of travel to walking, cycling, and public transportation."

13.4 Environmental Issues and the Policy (Economic Measures)

13.4.1 Environmental Issues and Transportation in Modern Society

Many environmental issues caused by transportation mainly include air pollution, global warming, noise, vibration, etc. In particular, the reduction of air pollutants and greenhouse gases is urgently needed in many countries. Although every mode of transportation generates some environmental impact, automobiles are specifically prioritized for regulatory and policy measures when assessed per unit of emissions. Air pollution directly affects human health, and materials such as NOx, SOx, PMx, etc., are generated locally. Although the emission of pollutant has been reduced compared to the past due to improvements in the performance of individual vehicles, in London, for example, Low Emission Zone has been established and charges have been made according to the performance of incoming vehicles. On the other hand, climate change is a phenomenon caused by a large amount of greenhouse gases, and the current global warming is particularly attributed to carbon dioxide. Carbon dioxide does not directly affect the human health, but its massive emission causes a rise in the earth's surface temperature, which is expected to have a significant impact in the future, such as sea-level rise and changes in vegetation and so on. So, this section will mainly discuss with a focus on the emission of carbon dioxide.

13.4.2 Understanding Environmental Issues from Economic Perspective

From an economic perspective, environmental issues are understood as the result of market failure. That is, as shown in Fig. 13.2 – with axes and curves identical to those in Fig. 13.1 – for activities that involve environmental external diseconomies, the excessive activity level X' is chosen according to the private marginal cost curve that does not include the relevant costs, and in such cases, environmental problems occur. In this case, deadweight loss is generated and social surplus is not maximized. As shown in Fig. 13.2, a graph illustrating economic concepts with three intersecting lines: the Marginal Social Cost Curve, Marginal Private Cost Curve, and Demand Curve. The x-axis represents quantity, marked with points X* and X', and the y-axis represents cost. Points A, B, E, and G are labeled on the graph, with a vertical distance labeled 't' between E and G. The graph demonstrates the relationship between social and private costs in economic theory. This forms the theoretical background for environmental taxes and carbon taxes. However, in the actual determination of tax levels, the Baumol-Oates tax is used in which tax rate is estimated and levied to achieve the scientifically derived allowable emission level, and the tax level is adjusted through trial and error basis.

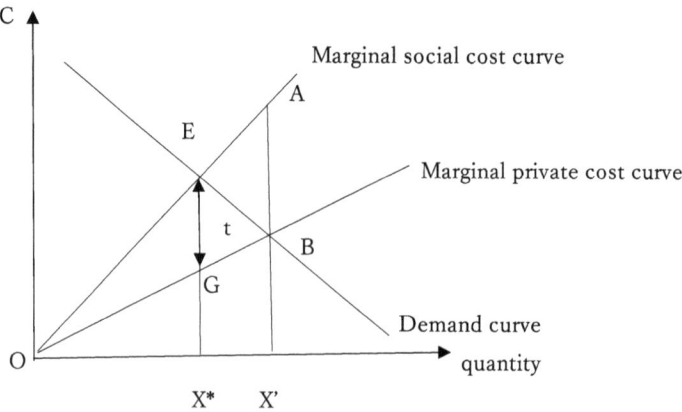

Fig. 13.2 An illustration of environmental external diseconomies

13.4.3 Measures to Address Environmental Issues

As indicated in the previous section, environmental issues arise from the excessive emission of the substances in question. The policy options for reducing emissions from automobiles can be classified into three categories: ① improving the performance of individual automobiles, ② reducing the use of automobiles, and ③ using automobiles efficiently. Respective measures are ① aims to reduce the emission unit of automobiles, ② aims to reduce the use of automobiles by switching to other modes of transportation, and ③ aims to reduce the number of automobiles in use by effectively utilizing automobile space.

Various policy innovations can be used to realize these three policies, but the use of "economic methods" that utilize market mechanisms can smoothly achieve the goals.

In the case of ①, tax reductions for acquisition and possession taxes for environmentally friendly automobiles are effective. Generally, consumers choose automobiles with better environmental performance and lower taxes, and automobile manufacturers are expected to strive to improve environmental performance. As a result, the average fuel efficiency of new automobiles will improve, and the average performance of owned automobiles will improve. For ② and ③, as shown in 13.4.2, efficiency at the usage stage can be achieved by adding to the fuel price.

13.4.4 Automotive Environmental Policy in Japan—Use of Economic Measures

In Japan, the "greening of automobile-related taxes" has been carried out since 2001 with the aim of improving the environment through improved automobile performance. This policy reduces acquisition and possession taxes for automobiles with excellent environmental performance, while imposing additional taxes on automobiles with poor environmental performance.

In response to this policy, automobile manufacturers, anticipating that consumers would choose low-pollution automobiles with lower taxes, made efforts to improve environmental performance. And the overall environmental performance of the automobile market improved, and the average fuel efficiency of owned automobiles also improved. As a result of these efforts, the amount of carbon dioxide emissions from the transportation sector in Japan is on a downward trend.

On the other hand, under the Paris Agreement, Japan has declared a 46% reduction by 2030 compared to fiscal 2013 and carbon neutrality by 2050. Measures centered on improving the performance of gasoline vehicles are insufficient to achieve this goal, and it is urgent to popularize next-generation vehicles such as HV, EV, and FCV. However, as of the end of March 2023, HV account for 18.5% of the owned automobile, while the proportion of EV is 0.26% and FCV is 0.01%. The reasons for the slow spread of next-generation vehicles include high vehicle prices and insufficient charging facilities and hydrogen stations. Also, from the perspective of promoting decarbonization, it is important to pay attention to the amount of carbon dioxide emissions in the manufacturing process of electricity and hydrogen, which are the energy used.

13.5 Balanced Growth or Unbalanced Growth

13.5.1 Balanced Growth Theory and Unbalanced Growth Theory

13.5.1.1 Nurkse's Balanced Growth Theory

The scale of a country's economy is determined by the three major production factors (land, private capital, and labor), savings rate, inflation rate, social overhead capital, technological level, and the terms of trade with other countries. Therefore, there are various analytical aspects in economic growth theory, and a research on growth paths is one of them.

Nurkse's balanced growth theory [7] is one perspective regarding the path that a country takes during economic growth Nurkse pointed out the lack of private investment as one case related to the failure of developing countries to realized economic growth. And Nurkse proposed balanced growth as a way of inducing private investment. In other words, he assumed that balanced production increases would expand markets. This theory is not supported in recent years.

13.5.1.2 Hirschman's Unbalanced Growth Theory

Hirschman advocated the theory of unbalanced growth [8]. Developing countries face various difficult-to-solve problems that impede economic growth, and such countries do not possess sufficient resources or capital to simultaneously solve them all. Therefore, as a strategy for rapid economic growth, Hirschman argued that the best strategy is to focus investment of resources on strategic growth industries and to maintain such an unbalanced state. Such strategic industrial growth would later have positive repercussion s for other industries and contribute to nationwide economic growth.

13.5.2 Social Overhead Capital and Private Capital

The unbalanced economic growth theory can be applied to the balance between private capital (private investment) and social overhead capital (public investment). In Fig. 13.3, a chart illustrating growth paths with axes labeled "Private capital" and

13 Sustainable Growth: An Economic Perspective

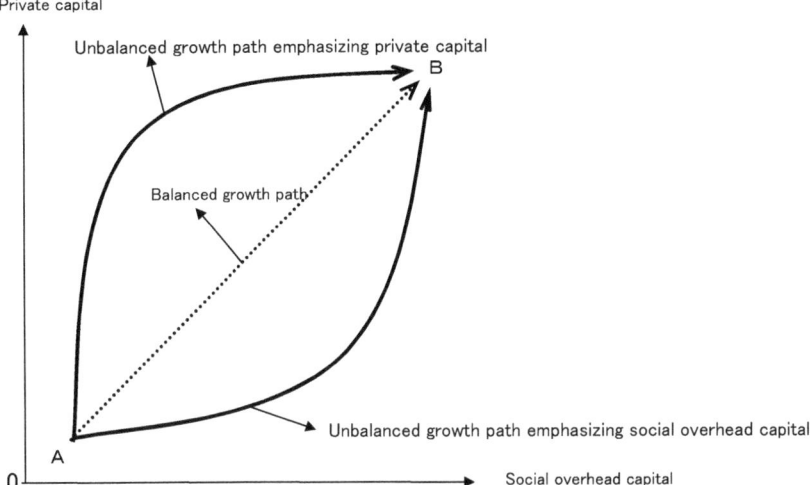

Fig. 13.3 Balanced and unbalanced growth paths

"Social overhead capital." Three paths are shown: "Unbalanced growth path emphasizing private capital," "Balanced growth path," and "Unbalanced growth path emphasizing social overhead capital." Points A and B mark the start and end of the paths. Arrows indicate direction of growth. The horizontal axis indicates the stock levels of social overhead capital, and the vertical axis represents the stock levels of private capital. The higher each stock level, the higher the potential productivity, and thus higher potential economic growth.

In Fig. 13.3, the dotted line at a 45-degree angle line indicates the path of balanced growth. In other words, investments occur in a balanced manner such that the marginal productivity is equalized between private and social overhead capital. In contrast, the curve at the top represents a growth path in which private capital takes precedence over social overhead capital, and the curve at the bottom conversely prioritizes social overhead capital development. The issue at hand is the speed and cost required to move from point A to point B. The theory of unbalanced growth states that an unbalanced growth path arrives at point B more quickly and at a lower cost (specifically, a lower opportunity cost of capital) than the balanced growth path does.

Yet when considering whether to prioritize private or social overhead capital, it is insufficient to consider only the opportunity cost of capital. This is because while the decision-making body for private capital investment is private and that for social overhead capital is primarily a public entity. Also, in the case of transportation infrastructure development, it is generally different in different regions within a country.

In cases where private economic growth is fast and social capital development cannot keep up, it will take the unbalanced growth path that emphasizes private capital. This would be demand-following social capital development in terms of transportation infrastructure development. On the other hand, in regions where private economic activities are stagnant, the government often intends to revitalize the region through transportation infrastructure development. This means that the government intends to adopt an unbalanced growth path that emphasizes social overhead capital. This is called development-leading social capital development. In this case, failure to attain a sufficient regional development effect would result in failure to reach point B, and it could be criticized as wasteful transportation infrastructure development.

13.5.3 Forward and Backward Linkage Effects

The economic effects of public investments such as transportation infrastructure construction are classified into two categories.

The forward linkage effects are the benefits (economic effects) from the public services provided by the facility developed through public investment. In other words, these are effects of increasing the stock of social overhead capital and enhancing potential productivity through public facility improvement. The forward linkage effects of a project are calculated as benefits of the project in CBA.

The backward linkage effects are the economic effects of increased demand for resources that are devoted to public investment, in particular economic effects on the regional economy. In other words, public investment activities result in increased demand for materials and labor, thus increasing incomes of material and labor suppliers. The backward linkage effects are often emphasized as being important economic effects of public investment programs for regional economies, especially in rural areas.

13.6 Transportations and Telecommunications

From the perspective of the transportation sector, telecommunications can be viewed economically in three ways: as a factor of production, as an alternative service to transportation, and as a complementary service to transportation.

13.6.1 Telecommunication as a Factor of Production

The main way to view telecommunications from the transportation sector is from the perspective of advanced transportation services utilizing telecommunication technology, such as smart mobility, which is discussed in another chapter of this book. From an economic perspective, this is a case in which telecommunication is a factor of production in the production of transportation services.

The development of telecommunication is basically an improvement in the efficiency of information transmission, so it allows us to share information that could only be obtained by actually going there and experiencing it, even without such action. For example, by sharing traffic congestion information in advance, it is possible to use the capacity-limited transportation infrastructure more efficiently. Also, in the case of cab services, there is an information asymmetry where the service quality of the driver is unknown in advance, but by sharing user ratings and other information using ICT (Information Communications Technology), it is possible to examine the quality of service in advance.

Some of the obstacles that have hindered smooth transportation service transactions are no longer a problem due to the development of telecommunication, and there are cases where institutional changes such as deregulation are possible.

13.6.2 Telecommunication as an Alternative Service to Transportation

As in the relationship between telecommunication and transportation, there are cases where the two services are in a substitutive relationship. In economics, if the price of one service increases, and the consumption of the other service relatively increases, the two services are defined as being in a substitutive goods relationship. A typical example of a substitutive relationship would be a case where the demand for travel decreases due to remote work. Friction against remote work, which was traditionally high, becomes easier as people get used to this form of work. The relationship between the lower cost (in terms of social friction) of implementing remote work and the resulting lower demand for transportation is a substitute relationship as defined by economics.

Communication has evolved from the era of letters to voice-only communication in the era of phone calls, with the addition of text communication such as e-mails, and the spread of video conferencing has made image-based communication possible. There is a possibility that we will continue to develop into more advanced communication in the future. It is highly likely that mobility will be increasingly replaced. However, the amount of information that can be obtained by traveling, communicating face-to-face, and experiencing the local area is enormous. A world where all this information circulates digitally is a long way off, and the transportation services that support the need for mobility are significant.

13.6.3 Telecommunication as a Complementary Service to Transportation

In the relationship between transportation and telecommunication, there are cases where the two services are complementary goods. In contrast to the case of substitute goods, the relationship of complementary goods is such that when the price of one service increases, the consumption of the other service relatively decreases. The fact that transportation services and telecommunication services are in a complementary relationship means that both services are used together as a set. For example, a case where a telecommunication network is established between two places, communication between the two regions increases, and so does travel by transportation. In fact, it is assumed that the current growth in various international passenger demands would not have proceeded without complementary communication provided by telecommunication services.

There are cases where the demand for travel is stimulated by the ease of obtaining information about travel destinations on the Internet. This example can be seen as an input of production factors by telecommunication services to the demand for travel, but viewing information about travel destinations by telecommunication services is a kind of virtual experience, and it is because of this that the decision to actually travel is made, so it can be said that telecommunication and transportation are in a complementary relationship.

13.6.4 Network Externality and Transition to New Services

Network externality has not been given much attention in the transportation field so far. Network externality refers to the effect of more people using the same service, thereby increasing the benefits to each user.

The various economic and psychological costs that consumers incur when switching to a new service are called switching costs, and due to the existence of network externality, the switching costs become large, making it difficult to transition to a new service, and this has been discussed so far.

For example, the global outbreak of covid-19 infection has rapidly popularized online meetings, but it is true that online meeting tools had existed before. The switching cost was high to reduce travel and actively use new online meeting tools, but because many people started using online meeting tools at once, the benefits of those tools dramatically increased due to the effect of network externality, and it can be seen that they have spread beyond the switching cost. Also, once the switching cost is overcome, it can be said that its convenience will continue to be widely used in the future.

The same thing is happening in the transportation field. For example, if all vehicles become autonomous vehicles, traffic congestion may be less likely to occur, but if only one vehicle becomes an autonomous vehicle and all other vehicles are driven by humans, it will not lead to traffic congestion relief. If many people do not switch to the new service at once and the network externality does not work, individuals may not be able to recognize sufficient benefits and may not be able to exceed the switching cost.

In cases where the spread of new services is socially desirable, public policies to encourage its diffusion may also be necessary.

References

1. Button K (2022) Transport economics, 4th edn. Edward Elgar
2. Mishan EJ, Quah E (2020) Cost-benefit analysis, 6th edn. Routledge
3. Kato H (ed) (2013) Theory and practice of time value of transportation. Gihodo Publishing. (in Japanese)
4. Brent RJ (1998) Cost-benefit analysis for developing countries. Edward Elgar
5. Dinwiddy CL, Teal FJ (1996) Principles of cost-benefit analysis for developing countries. Cambridge University Press
6. Transport for London. Congestion Charge marks 20 years of keeping London moving sustainably; 2023. https://tfl.gov.uk/info-for/media/press-releases/2023/february/congestion-charge-marks-20-yearsof-keeping-london-moving-sustainably
7. Nurkse R (1961) Patterns of trade and development. Blackwell. (mimeo 1959)
8. Hirschman AO (1958) The strategy of economic development. Yale University Press

Recommended Reading

Boardman A, Greenberg D, Vining A, Weimer D (2018) Cost-benefit analysis: concepts and practice, 5th edn. Cambridge University Press

Open Access This chapter is licensed under the terms of the Creative Commons Attribution-NonCommercial-NoDerivatives 4.0 International License (http://creativecommons.org/licenses/by-nc-nd/4.0/), which permits any noncommercial use, sharing, distribution and reproduction in any medium or format, as long as you give appropriate credit to the original author(s) and the source, provide a link to the Creative Commons license and indicate if you modified the licensed material. You do not have permission under this license to share adapted material derived from this chapter or parts of it.

The images or other third party material in this chapter are included in the chapter's Creative Commons license, unless indicated otherwise in a credit line to the material. If material is not included in the chapter's Creative Commons license and your intended use is not permitted by statutory regulation or exceeds the permitted use, you will need to obtain permission directly from the copyright holder.

Epilogue

In Conclusion: Towards Sustainable Transport
    Kazuhiko Takeuchi

Contribution to SDGs and Sustainable Transport

Outcomes of Rio+20 and Adoption of SDGs at the United Nations

How has sustainable transport been discussed in the international community? First, let's take an overview of the history of the SDGs.

In June 1992, the United Nations Conference on Environment and Development (UNCED), also known as the "Earth Summit," was held in Rio de Janeiro, Brazil. The Earth Summit was a conference aimed at obtaining concrete measures for the conservation of the global environment and the realization of sustainable development. At this summit, the Rio Declaration on Environment and Development, the Statement of Forest Principles, and Agenda 21 were adopted, and both the United Nations Framework Convention on Climate Change (UNFCCC) and the Convention on Biological Diversity (UNCBD) were opened for signatures. This marked the beginning of serious discussions on the conservation of the global environment.

Ten years later, in August–September 2002, the World Summit on Sustainable Development was held in Johannesburg, South Africa, to comprehensively review and discuss plans for implementing Agenda 21 adopted at the Earth Summit. Here, a focus was placed on five areas: water, energy, health, agriculture, and biodiversity, and the Johannesburg Declaration on Sustainable Development was adopted.

A further decade later, in June 2012, the United Nations Conference on Sustainable Development (Rio+20) was held again in Brazil, with discussions aiming at a future vision of the world, and a joint declaration called "The Future We Want" was adopted. Traditionally, discussions at UN conferences were conducted by delegations from member states, but one very significant feature of Rio+20 was that a diverse range of stakeholders participated in the discussions and formulated policies together with the member states. This approach of "stakeholders participating in discussions" was then carried over to the formulation of the Sustainable Development Goals (SDGs).

Structuring of 17 Goals and 169 Targets in SDGs

Consisting of 17 goals, 169 targets, and 232 indicators, the SDGs were adopted at the "United Nations Sustainable Development Summit" held at the United Nations Headquarters in New York in September 2015. SDGs are a successor to the MDGs (Millennium Development Goals) which were formulated in 2001.

The significant difference between the SDGs and the MDGs lies in the position of developed countries. Under the leadership of then Secretary-General Kofi Annan, MDGs consisting of eight goals, 21 targets, and 60 indicators were primarily aimed at eliminating the North-South divide. In this context, developed countries were expected to provide financial and technical support to developing countries. However, for the SDGs, the position of developed countries changed from supporters to stakeholders, with the aim of working together with developing countries to realize a sustainable world. These new global goals were set not only for developing countries but also for developed countries, and as such, SDGs are much more complex than MDGs, as evidenced by the numerous goals, targets, and indicators that support them.

In light of this complexity, the United Nations put forward the "Five Ps: People, Planet, Prosperity, Peace, and Partnership," as core pillars that form a framework for the 17 SDGs.

While developing countries had already established the MDGs, it was a major challenge to establish the SDGs in developed countries. It can be said that Japan is one of the countries that has a better understanding of the SDGs compared to other developed countries, and there has been a significant increase in the number of local governments and private companies that are proactive in working towards the realization of the global goals.

Local governments in Japan are making particular efforts in promoting sustainable city planning and regional revitalization initiatives. Such efforts are expected to have a synergistic effect of optimizing overall policies and accelerating the resolution of regional issues by incorporating the philosophy of the SDGs. In addition, local governments that propose exemplary SDGs initiatives are being selected as "SDGs Future Cities" with support given to pioneering initiatives including "Local Government SDGs Model Projects" (source: https://www.chisou.go.jp/tiiki/kankyo/miraitoshi.html).

In the case of the private sector, Corporate Social Responsibility (CSR) has promoted the establishment of the SDGs. As CSR has evolved, companies are required to ensure that their activities are related to social contribution, and thus the SDGs have been utilized very effectively.

The SDGs have several advantages, including the fact that they do not distinguish between developing and developed countries. Also, by combining a variety of goals and targets in a diverse way, initiatives related to the SDGs are possible not only at the national level but also at the local government and corporate levels. However, there are also challenges. One is that there are too many goals and targets. While it is easy to work on items included in the goals and targets of the SDGs, initiatives that are not included are likely to be left behind. The lack of legally-binding agreement is also a major issue in the international community.

Regarding the UN Framework Convention on Climate Change, the Kyoto Protocol was adopted in December 1997, and the Paris Agreement was adopted in 2015, establishing specific rules for the Convention. The Kyoto Protocol obligated each country to reduce greenhouse gas emissions, but the reduction targets for greenhouse gases set in the Paris Agreement for 2020 and beyond are not mandatory. They are merely reduction targets voluntarily set by each country. While the absence of penalties allowed developed and developing countries to reach an agreement, it has been pointed out that weak enforcement may result in countries not implementing their reduction targets.

Linking SDGs and Sustainable Transport

Considering these international trends, how should the SDGs be achieved in the transportation sector? One highly evaluated report was published in 2016 by the Secretary-General's High-Level Advisory Group on Sustainable Transport (Mobilizing Sustainable Transport) and included several policy recommendations, recognizing the importance of sustainable transportation in achieving the SDGs (Fig. 1).

Let's apply the goals and targets of the SDGs to the transportation sector. As shown in Fig. 1, in the transportation sector, for Goal 3 "Good Health and Well-Being for All," targets are set such as "halving the number of traffic accident deaths" and "significantly reducing the number of deaths from air pollution." For Goal 2 "Zero Hunger," infrastructure issues in developing countries are pointed out. Due to the lack of access to markets, agricultural products produced in developing countries do not reach the market while they are still fresh, resulting in waste. For Goal 7 "Affordable and Clean Energy," energy efficiency and the overuse of fossil fuels are seen as challenges. For Goal 11 "Sustainable Cities and Communities," it can be difficult to incorporate climate change policies into national policies and build a sustainable transportation system that is safe, affordable, and accessible. Goal 13 "Climate Action" requires the development of sustainable and resilient infrastructure.

In Japanese society, the idea of systematically representing issues in the transportation sector is not well established. Therefore, at the International Association of Traffic and Safety Sciences (IATSS), we believe it is necessary to expand our activities beyond SDGs and Sustainable Transport.

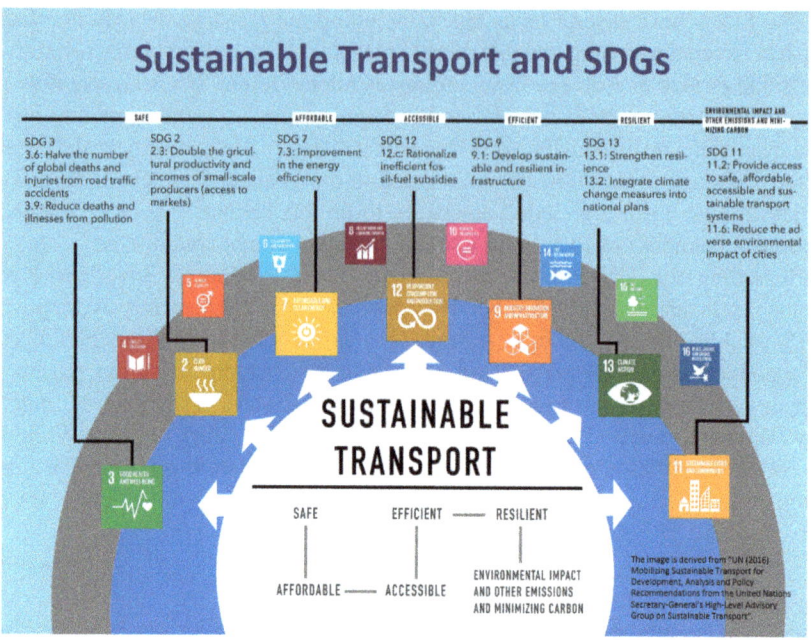

Fig. 1 Based on mobilizing sustainable transport for development (UN Secretary-General's High-Level Advisory Group on Sustainable Transport, 2016)

The Development of Traffic and Safety Sciences from the Perspective of IATSS's 50-Year History

IATSS Activities Beginning with the Traffic War and Oil Crisis

The International Association of Traffic and Safety Sciences (IATSS), a public interest incorporated foundation, was established by Soichiro Honda and Takeo Fujisawa, the founders of Honda Motor Co., Ltd. Here, I would like to take a moment to look back on the history of the establishment of IATSS.

The issue that both founders were concerned about was the "traffic war." In line with economic growth since 1955, the number of traffic accident fatalities in Japan had increased significantly. In 1970, the number of traffic accident deaths was 16,765, a number comparable to the 17,282 people who died in the Sino-Japanese War over two years. This situation, comparable to a state of war, was named the "traffic war." The 1960s saw the arrival of what was called the first traffic war, but the number of traffic accident fatalities decreased thereafter. However, in the 1980s, fatalities began to increase again, and this period is referred to as the second traffic war.

In addition to the traffic war, the 1960s was also a time when the problem of pollution, such as automobile exhaust emissions, was becoming serious. In order to solve these problems in a systematic and integrated manner, it was thought necessary to conduct interdisciplinary research that transcends individual fields of expertise, rather than rely on experts in each field independently addressing the problem.

Moreover, there were limits to what private companies could do to tackle these social issues on their own. Therefore, in September 1974, Soichiro Honda and Takeo Fujisawa established IATSS using their personal funds.

IATSS has been focusing on two main research subjects since its inception: traffic safety and the environment that supports traffic, with the aim of "contributing to the realization of an ideal mobile society." One of the characteristics of IATSS is that it only has 59 members, as of August 2024. The intention is to foster an interdisciplinary culture by gathering a very small number of outstanding experts with an emphasis on diversity between fields. The majority of members are researchers in traffic engineering, but researchers from a wide range of fields such as medicine, psychology, sociology, environmental studies, as well as people from outside the research community, such as cultural figures, also participate in IATSS activities. With these characteristics, IATSS makes a significant contribution to advancing comprehensive research on a sustainable traffic society.

Evolution from Interdisciplinary Research to Transdisciplinary Research

One major global goal of IATSS is the transition by 2050 to a so-called carbon-neutral society, which essentially reduces greenhouse gas emissions to zero by balancing them with absorption or removal. One significant challenge is how far emissions can be reduced by 2030. There are certainly considerable hurdles to reducing greenhouse gas emissions to virtually zero. It is not enough for experts in various fields to produce academic research results in terms of social effects, rather it is important for various stakeholders to participate from the early stages of research and for experts and stakeholders to think together about how research is conducted. Traditionally, an approach has been taken whereby each expert works on problem-solving in their respective fields. This is called "multidisciplinarity." In contrast, an approach where experts from different fields collaborate to produce a single result is called "interdisciplinarity." Furthermore, an approach to problem-solving as a society as a whole with various stakeholders is called "transdisciplinarity." It is important for IATSS activities to expand from interdisciplinary research to transdisciplinary research (Fig. 2).

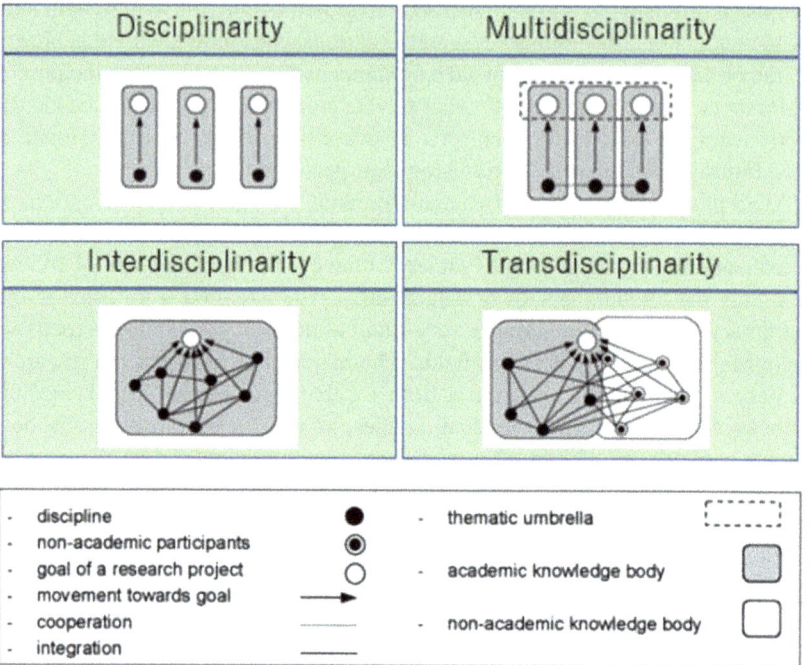

Fig. 2 Tress, G., Tress, B., Fry, G. (2004): Clarifying integrative research concepts in landscape ecology. *Landscape Ecology*, 20: 479–493

Collaboration with IATSS Forum Projects

The year 2025 marks the 50th anniversary of IATSS. Furthermore, the IATSS Forum, which was established 10 years after IATSS in 1985, is celebrating its 40th anniversary. The Forum was also established by Soichiro Honda and Takeo Fujisawa using their personal funds, just like IATSS. I would like to touch on the background of the establishment of the IATSS Forum and its future development direction. In 1981, the "Look East Policy" was proposed by Mahathir bin Mohamad, the fourth prime minister of Malaysia. This policy, also known as the Eastern Policy, aimed at economic development in Malaysia by learning from the strong work ethic, work motivation, and management skills that were observed in Japan. Prime Minister Mahathir stated that "the future of Asia lies in human resources," and he highly valued human resource development. This was something that Honda and Fujisawa also strongly supported and thus they established the IATSS Forum as an education program. The current venue for the IATSS Forum is located in Yokkaichi City, Mie Prefecture, but for a long time after its establishment, it was held on the premises of the motorsport racing track, Suzuka Circuit. The Forum is composed of participants who are young professionals in ASEAN countries and India, including researchers, public officials, people working in private companies, entrepreneurs, etc., with about 20 people participating. Although the parent body is IATSS, the Forum focuses on human resource development, with a rather noticeable lack of collaboration with the IATSS main body. Therefore, to strengthen the collaboration between the two, the curriculum of the IATSS Forum is currently being reviewed.

One of the pillars of the IATSS Forum is to nurture human resources committed to solving transportation problems, environmental issues, and regional challenges not only in Japan but also across the world. Yokkaichi City, where the IATSS Forum is located, is a city that once had major problems with air pollution, and the respiratory diseases it caused were known as "Yokkaichi Asthma." Holding the IATSS Forum in Yokkaichi, a city that has overcome pollution, is considered beneficial for trainees from developing Asian countries that still face pollution problems. By linking the IATSS Forum with global issues such as the SDGs, we hope to nurture individuals with insight into sustainable development and sustainable transportation systems in various parts of Asia.

It is also important for participants at the IATSS Forum to continue to interact continuously and voluntarily, rather than ending their experience as a one-time training. Therefore, the IATSS Forum holds alumni meetings in various countries, and many graduates continue to be actively involved. From a global perspective, ASEAN countries and India are still plagued by serious traffic problems. For example, Vietnam is a country where motorcycles dominate the roads, and as a result, there are many motorcycle accidents. Honda is a market leader with a significant presence in the region, and as such, the IATSS Forum has the potential to make a strong and meaningful contribution to traffic safety in Asia.

The Future for a Transportation Society and Lifestyles in Japan

Transportation Society in the Era of Decarbonization and Coexistence with Nature

Currently, we are probably experiencing the greatest transition since the Industrial Revolution, which was when our dependence on fossil fuels began. Global companies are taking action in light of this transition. For example, Nippon Steel has formulated a "Carbon Neutral Vision 2050," declaring that it will reduce its CO_2 emissions, including those from its supply chain, to virtually zero by 2050. When aiming for carbon neutrality, there needs to be a transformation of the structure of the blast furnace itself. Similarly, in the transition from vehicles running on conventional engines to electric vehicles, we cannot respond to the change without altering the energy supply system. At present, the parking spaces at our homes are merely places to park cars, but if electric vehicles become widespread, cars will be used as batteries, and parking lots will take on the role of supplying electricity to cars. In this way, society itself will likely change significantly as cars change.

Another current problem is that the extinction of biological species is accelerating worldwide. In order to create a society where we humans can coexist with nature, the term "Nature Positive" has become established in discussions triggered by the 15th Conference of the Parties to the United Nations Convention on Biological Diversity (CBD COP15). This is an attempt to reverse the degradation of natural capital, which is on a downward trend, by 2030. From the perspective of transportation, contributions can also be made through greening of roads, embankments, interchanges, etc. It is also necessary to consider a transportation society from a nature positive perspective.

Transportation Society in an Era of Population Decline and Aging

Rapid population decline and aging are progressing not only in Japan but also in South Korea and China. Population decline is particularly noticeable in South Korea, which has a birth rate of 0.78, compared to birth rates of 1.26 in Japan, and 1.15 in China. In terms of the aging rate of those aged 65 and over, Japan is at 37.7%, China at 30.1%, and South Korea at 37.0%.

In Japan, population decline and aging is an especially serious issue in so-called mountainous areas, which are often depopulated areas where transportation has become a major problem. In these mountainous areas, private cars are a vital part of life. Therefore, even in their 70s and 80s, local residents use private cars for transportation. Although it would be preferable to develop public transportation in these areas, this is costly and requires personnel for maintenance, so it is necessary to consider how to switch public transportation to a form suitable for an aging society.

Here, I would like to introduce initiatives taken by Yamato Transport Co., Ltd. and Northern Iwate Transportation Inc., two companies which won the 37th annual International Traffic Safety Association Award in 2015 presented by IATSS. These two companies have created mixed passenger and cargo transportation on two routes: the Morioka-Miyako intercity bus route and the Miyako-Omoe Peninsula local bus route. By modifying the bus vehicles of Iwate Kita Motor Co., Ltd. and loading both passengers and cargo onto the bus, both human mobility and logistics have been maintained.

Another city that is taking innovative initiatives is Toyama City, which was the first city in Japan to launch a compact city policy. According to the Aging Society White Paper published in 2019, this policy encompassed "a compact city development centered on public transportation by activating public transportation, including railways, and concentrating various urban functions such as residence, commerce, business, and culture along the public transportation lines." The policy is characterized by "freely choosing the place of residence, comprehensively enhancing the attractiveness of the city center and public transportation lines, and gently guiding the residence and urban functions to prevent the diffusion of the city area and ensure the quality of life and sustainability of the city." By introducing LRT (a next-generation tram system), Toyama aims to revitalize public transportation and create a convenient community that is both elderly- and child-friendly.

In the future, autonomous vehicles will probably become established in society and have the potential to solve challenges related to the declining birthrate and aging population. In any case, modes of transportation will continue to change and indeed must change, with this transformation supported by science and technology. In addition to an autonomous driving system, we have seen recent examples of goods being delivered by drones, which also have been put to use spreading fertilizer and pesticides on fields. In fact, it will be possible to reduce the use of pesticides by having the drones measure the air and then spray pesticides only where necessary. In this way, new technologies will increasingly be used in local communities.

Growing Asian Transportation Society and International Cooperation

Simultaneous Resolution of Pollution and Global Environmental Issues

Initial research activities by IATSS have been centered in Japan, but in recent years, activities have been expanding internationally, with projects in other Asian countries such as India and Cambodia. In Thailand, an organization related to IATSS called ATRANS is also active. ASEAN countries and India are also regions with many IATSS forum trainees. These regions are experiencing rapid population growth and are plagued by traffic pollution, much like Japan in the past.

From the late 1960s to the early 1970s, the number of traffic accident fatalities in Japan surged, and pollution from exhaust gases was also severe during this period. After resolving its own pollution issues, Japan then faced global environmental problems. However, in ASEAN countries and India, the impacts of climate change and air pollution are occurring and must be dealt with simultaneously. The approach to achieving both greenhouse gas reduction and air pollution prevention at the same time is called the "co-benefit approach," and this is exactly what is required in ASEAN countries and India. Cities across the region continue to grow, but there is also a need to consider the nature of new transportation systems that can keep up with the speed of growth. In the past, traffic congestion in Bangkok, Thailand, was extremely serious, and it was not unusual to take two or three hours to reach a place that would normally take 30 minutes. As a countermeasure, the BTS (elevated railway) started running in 1999, significantly alleviating congestion. On the other hand, traffic congestion has not been resolved in Jakarta, the capital of Indonesia. Here, about half of the total population is concentrated on the island of Java, where Jakarta is located, and emissions of greenhouse gases and air pollution due to traffic congestion are becoming very serious problems. Therefore, the government is promoting the relocation of the capital to a new city called Nusantara on the island of Kalimantan, a proposal that has been under consideration for some time. Initiatives such as these also come under a co-benefit approach. Discussions should be promoted so that Japan's experience can be utilized to deal with transportation problems in other Asian countries. To do this, it is essential to transform the IATSS Forum, thereby bringing more opportunities to deepen understanding of Japanese transportation systems, and to organically link the research activities of IATSS itself and initiatives carried out by the IATSS Forum. As the region heads to a period of great transformation, IATSS and the Forum could collaborate to help alleviate the problems faced by Asian countries. IATSS must continue to uphold the philosophy of contributing to the realization of an ideal transportation society, but a different approach may also be required. In an ideal transportation society, there are two approaches: reducing negative impacts to zero, and realizing positive impacts. For example, an approach to reduce negative impacts to zero could be to reduce the number of traffic accident fatalities to zero. An approach to realize positive impacts could be, for example, to create a transportation system that is easy for an aging population to navigate. It could be said that an ideal transportation system is one in which transportation contributes to the improvement of people's well-being.

Issues and environmental problems related to transportation are subject to change as time passes. We need to think about new activities that can flexibly respond to change and that can adapt in periods of transformation.

Co-evolution of Interdisciplinary Research in Japan and Other Asian Countries

Traditionally, the mainstream idea was to transfer technologies established in developed countries to developing countries, but this idea is no longer valid. From now on, we need to carry out research activities based on co-benefits and co-production (joint creation), where problem-solving approaches in developing countries also lead to problem-solving in developed countries. As we head into the future, it will be crucial to formulate a method of co-evolution that not only conveys Japan's experience to ASEAN countries and India, but also advances interdisciplinary research between Japan and Asian countries with stakeholders and researchers alike.

The Future of IATSS Supported by the Next Generation

Going forward, one very important issue is how to nurture young talented professionals. In Japan, the retirement age has gradually been raised from 55 to 60, and further to 65. This is a requirement for an aging society, but it should not close the possibility for young people to participate fully in society. From this perspective, IATSS is aiming to slightly increase its membership and open the door to young talented scholars. IATSS was established with donations from the founders of Honda Motor Co., Ltd., a company that Japan takes pride in globally. I would like to conclude by expressing my gratitude to Soichiro Honda, Takeo Fujisawa, and Honda Motor Co., Ltd., as it is undoubtedly thanks to them that IATSS has reached its 50th anniversary and thus hope remains of realizing an ideal and sustainable transportation system.

GPSR Compliance
The European Union's (EU) General Product Safety Regulation (GPSR) is a set of rules that requires consumer products to be safe and our obligations to ensure this.

If you have any concerns about our products, you can contact us on

ProductSafety@springernature.com

In case Publisher is established outside the EU, the EU authorized representative is:

Springer Nature Customer Service Center GmbH
Europaplatz 3
69115 Heidelberg, Germany

www.ingramcontent.com/pod-product-compliance
Ingram Content Group UK Ltd.
Pitfield, Milton Keynes, MK11 3LW, UK
UKHW021825230326
469280UK00001B/10